Lightning in the Andes and Mesoamerica

Lightning in the Andes and Mesoamerica

PRE-COLUMBIAN, COLONIAL, AND CONTEMPORARY PERSPECTIVES

John E. Staller and Brian Stross

OXFORD
UNIVERSITY PRESS

OXFORD
UNIVERSITY PRESS

Oxford University Press is a department of the
University of Oxford. It furthers the University's objective
of excellence in research, scholarship, and education
by publishing worldwide.

Oxford New York
Auckland Cape Town Dar es Salaam Hong Kong Karachi
Kuala Lumpur Madrid Melbourne Mexico City Nairobi
New Delhi Shanghai Taipei Toronto

With offices in
Argentina Austria Brazil Chile Czech Republic France Greece
Guatemala Hungary Italy Japan Poland Portugal Singapore
South Korea Switzerland Thailand Turkey Ukraine Vietnam

Oxford is a registered trade mark of Oxford University Press
in the UK and certain other countries

Published in the United States of America by
Oxford University Press
198 Madison Avenue, New York, New York 10016

Library of Congress Cataloging-in-Publication Data
Lightning in the Andes and Mesoamerica :
pre-columbian, colonial, and contemporary perspectives /
John E. Staller and Brian Stross.
p. cm.
ISBN 978-0-19-996775-9
1. Lightning—Latin America—Folklore. 2. Lightning—Religious aspects—Latin America.
3. Indian cosmology. Indians—Religion. 5. Indians—Rites and ceremonies.
6. Latin America—Antiquities. I. Staller, John E. II. Stross, Brian.
GR630.L54 2013
299.8'113—dc23 2012027303

1 3 5 7 9 8 6 4 2

Printed in the United States of America
on acid-free paper

CONTENTS

PREFACE

In the pre-Columbian Andes and Mesoamerica the three most powerful and venerated celestial deities were lightning, the sun, and the moon. In the Inca seat of royal power, Cuzco, a faceless golden idol of lightning was kept with idols of the sun and of the moon, in the Coricancha or Golden Enclosure, the most sacred temple in the imperial capital. That peoples and cultures in these regions, separated by more than two thousand miles of ocean and mountainous terrain, should focus their veneration on lightning, the sun, and the moon is perhaps not so surprising. These were agricultural civilizations that needed to keep track of solar and lunar cycles, along with the coming rains and dry seasons, in order to survive. As a result, the pre-Columbian cultures of these culture areas created highly sophisticated and detailed calendars that kept track of solar and lunar cycles as well as the cycles of Venus. Some scholars speculate that hieroglyphic writing in Mesoamerica and *quipus* (knotted cords) in the Andes were created in these culture areas in order to provide a means of keeping track of annual cycles, resources, and managing their polities. The sun, with its steady progress along the plane of the ecliptic from sunup to sundown, creates a daily timetable, periodic warmth, and light to see by, in addition to providing the energy for the photosynthetic redistribution process that underlies most life on earth. The moon and its lunar cycles are tied to the night sky and thus represent the celestial complement to the sun, its symbolic associations more directly tied to the Other World or *mundo interior* (inside world), and to the reproductive cycles of a number of species of plants and animals. The moon controls the ocean tides and has associations to the female menstrual cycle and thus to blood. Through such associations the moon is commonly linked to female gender and perceived by many indigenous cultures to control important aspects of agriculture and human fertility. Lightning is perhaps the most awe-inspiring of the trinity: certainly the most feared, in effect the most capricious, and often understood to be the most immediately powerful. Lightning can strike out of the blue, kill in an instant, and create raging fire, and the associated rain and hail can create floods and destroy agricultural crops. Yet, like the sun and the moon, lightning is often perceived as a bringer of fertility. Its common occurrence with the coming rains, so necessary for crops, and its production of nitrogen that can be utilized by vegetation, explain at least in part its importance to the fertility and fecundity of the earth. We were both more interested in lightning and its place in the cultures and imaginations of people than in the sun and the moon. The reasons for our interest should become clear in this book, and we hope to have provided some data and some stimuli for further studies of lightning. Despite a wealth of scholarly and scientific information regarding the sun and moon and their associated cycles to the world's civilizations, and despite lightning's iconic importance to ancient civilizations throughout the world, lightning is the least understood and researched—particularly with regard to its significance and importance to ancient religious ideologies, folklore, and legend.

Our coauthored book came out of a series of collaborations beginning in 2004, when Brian Stross contributed chapters to several books that were edited or coedited by John Staller. During the course of our communications we came to realize that we both had independently wondered about lightning with regard to the worldviews and cosmologies of indigenous peoples in these culture areas. Staller has long had an interest in lightning in the Andean region, going back to his dissertation research. He has published on the possible existence of archaic forms of lightning veneration that may have created ethnogenesis or cultural horizons, and has long had an interest in how such horizons and the spread of politico/religious ideologies are associated with language. Stross, a linguistic anthropologist by training, had been intrigued by the apparent lightning connections of Classic Maya predecessors of both God B (called the rain god) and God K (called by some the god of dynasty, but recognized also as a lightning god) of the Maya codices, and fascinated by the fact that while animal co-essences or "familiars" are attributed to both humans and the lightning deity in Mesoamerica, some people in contemporary cultures also have lightning as a co-essence. On discovering our mutual interest in lightning, Staller suggested that we coauthor an article on the symbolic associations of lightning in the Andes and Mesoamerica, and thus what came to be a book project on lightning was initiated.

Stross initially produced a list of sixteen cultural and symbolic associations that could be found for lightning in much of Mesoamerica past and present, beginning with ethnographic and language-related information, realizing in the process that these associations not only tell something about how lightning is perceived and imagined, but also that, even without pursuing these associations any further, they present a fascinating picture (if only a specialized and partial one) of religious thought and practice in these ethnically and culturally distinct regions of the New World. Staller found a number of similar associations with lightning in the Andes and with his scholarly interest in archaeology and Colonial accounts of the indigenous population, focused on these aspects of lightning perceptions for the Andes. Our earlier drafts produced rather different presentations of the separate regions, reflecting not only the different regions themselves and the different histories that they have endured, but also differences in our training, biases, and presentation styles, in addition to differences in what we considered meaningful and relevant to the pre-Columbian, Colonial, and contemporary populations in these culture areas. We value these differences because they provide some information about the filters through which the data have been drawn in each case, as well as the different perspectives that we brought to bear on the original question of what the cultural perceptions and lightning associations were in these two ecologically, culturally, and geographically distinct regions of the New World.

Stross went through the Andean and Mesoamerican sections, initiating a discussion and compiling the similarities and differences to which Staller added a great deal of relevant information. The conclusion was similarly generated, and as we sent the manuscript back and forth, we realized that the article we initially planned had become a book. Although each author has had primary responsibility for the various lines of evidence concerning his own culture area, as a whole it has been truly a collaborative effort. We have reason to thank each other for engaging in this collaboration with a minimum of the kind of stresses that sometimes present themselves in a joint undertaking of this sort.

The book was sent out for publisher review, and the reviewers provided some useful suggestions for various ways of improving it. We were sincerely grateful for the constructive comments of several anonymous reviewers. We also extend our sincere thanks to Dr. Paul Heggarty (Max Planck Institute for Evolutionary Anthropology, Leipzig) and Dr. David Beresford-Jones (McDonald Institute for Archaeological Research, University of Cambridge, UK) for kindly passing along their generous comments and many insights regarding the processes underlying language spread in the Andes. In their careful readings of earlier drafts of this manuscript, they provided both encouragement and valuable information on how we had to proceed with this comparative study. We are greatly indebted to both. Thanks also to Dr. Tristan Platt (University of St. Andrews, Scotland, UK) for kindly passing along some of his published research and that of his colleagues and their research around Potosi, Bolivia. We are additionally indebted to Xmal Ton, Alonso Mendez, Santiago Tzapat, and Slus Bot of Chiapas, Mexico, for multiple ethnographic insights, and to linguists Rosemary Beam de Azcona and Mark Sicoli for generously providing unpublished data gleaned from their linguistic and ethnographic fieldwork in Oaxaca. All interpretations and statements of fact are solely our responsibility.

Lightning in the Andes and Mesoamerica

1

Introduction

Lightning has evoked a numinous response as well as powerful timeless references and symbols among ancient religions throughout the world. The thunderbolt is a conspicuous image among the ancient civilizations, and thunder and lightning also have very powerful associations with some of the primary deities of the ancient world (**Figure** 1). The thunderbolt image is present in ancient Mesopotamia, associated with Sumer, Babylonia, Akkad, and Assyria, and later inherited by the Greco-Roman tradition[1] through Zeus and Jupiter respectively (**Figure** 2). Zeus is frequently depicted in one of two poses: seated in majesty or standing, striding forward, with a thunderbolt leveled in his raised left hand (Bulfinch 1976:4). Jupiter or Jove essentially represents the Roman counterpart of Zeus, with similar symbolic referents, and he is likewise the primary deity in the pantheon (Bulfinch 1976:40). In ancient Rome, places where lightning struck were "considered holy and became the sites of oracles and temples" (Friedman 2008:30). These oracles and temples were consecrated to Jupiter, deity of the sky, and of thunder and lightning. Lightning divination was practiced some three hundred years before and after the current era and had definite political implications. The Romans believed that a lightning bolt passing from left to right was a favorable omen, while a bolt passing from right to left was a sign that Jupiter disapproved of current political events (Friedman 2008:31). Thor is the god of thunder and lightning in Norse myth and cosmology (**Figure** 3). Thor wielded a hammer and is mentioned prominently throughout the recorded history of the Germanic peoples. The name Thor is from Old Norse *Þórr*. Linguists indicate that the Old English *Þunor* and High German *Donar* are cognates within Germanic languages descending from a Common Germanic word, **þonaroz* or **þunraz*, which means "thunder." The

[1] Zeus ("living one," from the Sanskrit *dyaus*, "heaven") is the primary deity of Greek mythology and is associated with the sky and thunder.

FIGURE 1 The Mesopotamian sun god Marduk, patron of Babylon, holding the thunderbolt (*vajra*) in both hands pursuing the Chaos Monster, Anzu. The *vajra* was a symbol of the thunderbolt among the various Indo-European civilizations. This carved stone panel is from the ancient Assyrian site of Nineveh, located along the Tigris River in what is today northern Iraq, with occupations dated to c. 1800 BCE (from Layard 1853: Plate 19/83).

Celtic god of thunder, Taranis, and the Irish god Tuireann are also symbolically related. Norse mythology, largely derived from Scandinavia, has numerous legends and tales associated with this deity (Bulfinch 1976:331–332, 355).

In the Hindu and Buddhist[2] tradition, Indra is the storm-god, and as the bringer of rain he is commonly invoked throughout the Indian countryside (Ions 1967:15). Indra is often depicted holding the thunderbolt Vajra in his right hand. In Hindu mythology, Indra is the king of the gods or *devas*,[3] Lord of Heaven, god of storms, rainfall, and war. Indra hurls thunderbolts using a rainbow as a bow (Ions 1967:73). Post-Vedic texts, the Upanishads, speak of a great monster, Vritra, that enclosed all of the waters of the world, bringing on a terrible drought (Ions 1967:15). Indra seized a thunderbolt and dropped it on

---

[2] The Aryans entered the Indus Valley at around 1700 BCE, and narratives of their gods stem from the Vedas, a collection of hymns completed around 800 BCE (Ions 1967:14). The mythology and legends associated with the Vedas have come to be seen as *apauruseya* (of divine origin), revealed indirectly by what is *sruti*, or "heard," and *smrti*, "remembered" (Apt 1965:109).

[3] Indra is the Sanskrit term (इन्द in Devangari, the *abugida* alphabet of India and Nepal), and *deva* is the Sanskrit word for deity and can be loosely interpreted as any supernatural being (Apt 1965).

FIGURE 2 Zeus of the sky and thunder, the primary deity of Greek mythology. In this pose, Zeus is standing with a thunderbolt leveled in his raised right hand. Zeus represents the counterpart of Jupiter or Jove in the Roman pantheon. This statue is known as *Zeus of Smyrna*. Found in 1670 in İzmir, Turkey, it was restored by Pierre Granier, c. 1686. The arm raising the thunderbolt was added in restoration. Marble, middle second century, the Louvre, Paris.

FIGURE 3 Thor wielding his hammer. A primary deity in Norse mythology, Thor was the god of thunder and lightning and was associated with the power of nature and death. (*Thor's Battle with the Ettins*, by Márten Eskil Winge, oil on canvas, 1872.

the monster, blowing him up and thus releasing the waters (Ions 1967:15–16). Indra then climbed Mt. Meru, the cosmic mountain, and had a magnificent palace built in his own honor (Ions 1967:73; Campbell 1974).

Indra is believed to regulate the heavens as well as the annual seasonal cycle (Ions 1967:16). Indra is the chief *dharmapala*, or wrathful deity, of Hindu mythology. He is defender and protector of the Buddha, Dharma, and Sangha (Ions 1967:73, 76; Campbell 1974:169).[4] The thunderbolt or *vajra* is one of the most powerful symbols of Buddhist iconography, representing indestructible enlightenment (Campbell 1949:87–88). The Adi Buddha represents the Tibetan Vajra-Dhara (Tibetan: Dorje-Chang) or holder of the brilliant lightning bolt (Campbell 1949:88; 1974). Throughout the ancient world, thunder and lightning had primary importance and associations among the various pantheons, religious ideology, and origin mythologies with common symbolic references to death, rainbows, the coming rains, caves, mountain summits, and war. Similarly, lightning veneration played an important role in the ancient civilizations of the Andes and Mesoamerica. Lightning veneration and the religious cults and their associated rituals represent, to varying degrees, a worship of nature and the forces that shape the natural world.

The interrelatedness of the cultural and natural environment stems from what may be called a widespread cultural perception of the natural world as sacred—a kind of mythic landscape (Moore 1984; Lane 2002; Bauer 1998; Moore 2005; Staller 2008a). Comparative analysis of the Andes and Mesoamerica has been a recurring theme ever since Kirchhoff (1968) attempted to delineate culture areas in the Americas and to define them in terms of various culture traits (see also Armillas 1957, 1968). It has been a recurring theme in part because two of the areas of "high civilization" in the Americas have much in common despite substantial ecological differences, and in part because there is some evidence, of varying quality, that some people have migrated from one area to the other. There is other evidence that some items, particularly food crops, have found their way from one region to the other, leading to the important questions: How many and which of the apparent commonalities found between the two culture areas is due to independent innovation and convergence, and how much to pre-Columbian migration and/or diffusion? Was there trade between the regions, and if so, when did it occur? Was it through indirect contact, direct contact, or both, and was Central America from Honduras to Colombia involved in any such interactions? Which commonalities might be ascribed to colonial influence?

---

[4] Sometimes Indra is called Śakra and embodies the power of all primordial Dhyani Buddhas among the Buddhists. Indra also has symbolic reference to the ancient Zoroastrian tradition and to traditional religions in Burma, Thailand, China, and Japan. In Malay he is called Indera, and in Tamil, Intiran (Apt 1965; Ions 1967).

Our comparison, while bringing to the foreground some of the commonalities that have not been identified before and thus allowing for an analysis of whether or not they might have independent roots, cannot be expected to resolve these basic questions about the nature of the relationship between the Andean region and Mesoamerica in past times, since it involves only lightning and directly associated concepts, but it does sort through some of the more important commonalities and differences with respect to lightning veneration. We are particularly focused upon native perceptions of how lightning fits into individual and regional cultural traditions, allowing at least for an assessment of whether or not similarities of these sorts could have arisen independently or, alternatively, might have required some explanation in terms of diffusion or stimulus diffusion. Our goals regarding comparative analysis are thus not only to separately present the cultural webs of relationships binding the perceptions of lightning to other related phenomena and cultural traditions through time in both Mesoamerica and the Andes, but also to evaluate how various associations held in common might have come to be shared. Beyond that we seek also to understand the specific forms and contexts of lightning associations that are not held in common by inhabitants of the two regions.

Enough is already known about each region and of possible connections between the Andes and Mesoamerica to allow for some interesting comparisons and conclusions. We maintain the additional goal of more fully understanding how natural phenomena associated with lightning functioned in indigenous religions in these two culture areas in order to refine our knowledge of how deities were and are conceptualized and of indigenous worldview in general. This process should allow for the revision of simple analogies with the pantheon of classical deities, whose natures are also probably obscured by use of this label.

Our interdisciplinary research extends from pre-Columbian cultures to colonial clerics, as well as to current folklore and legends among indigenous populations that have reference to lightning. Our data provide evidence of acculturation and document the varying degrees to which native societies have maintained their earlier meanings or have otherwise interacted with colonial entities to produce syncretic cultures, influenced in particular by Western religious orthodoxy. Lightning veneration was of particular concern to early colonial Spanish clerics in these regions because the patron saint, Santiago or St. James, was associated with the thunderbolt and believed by them to have had a major role in the conquest of the New World (Madariaga 1947). Chroniclers indicate that Spanish clerics emphasized symbols, and associated religious beliefs regarding traditional lightning veneration to the cult of Santiago or St. James, the patron saint of New Spain (Krappe 1932; Madariaga 1947; Cardinale 1983). Ethnographic accounts and ethnohistoric documents provide insights through firsthand accounts that broaden our understanding of levels of syncretism since the European contact. Indigenous cultures in the Andes and

Mesoamerica transferred their veneration from the pre-Columbian deities to Catholic saints through conversion or the fusing of indigenous perceptions to Catholic orthodoxy, and, with respect to lightning, particularly to the cult of Santiago (Cardinale 1983; Topic et al. 2002). Among contemporary Andeans, syncretism of lightning involves fusing the ancient concepts with the Catholic saints Santiago or St. James and to a lesser extent Santa Barbara, patron saint of miners (Gade 1983).[5] In Mesoamerica, San Juan is more frequently linked to the lightning deity than is St. James (Santiago). The folklore, legend, and cultural beliefs surrounding lightning are to varying degrees an indirect reflection of how later Western religions were absorbed and integrated into indigenous and Latino ethnic and cultural identities (Staller 2010b).

This volume explores the symbolic elements surrounding lightning, as they existed in the associated pre-Columbian religious ideologies and iconography. Our research extends to contemporary cultures to reveal how cultural perceptions of the sacred, their symbolic representations, ritual practices, and architectural representations in the landscape were conjoined in the ancient past. It provides a basis for tracing back pre-Columbian manifestations of lightning and its associated religious beliefs and mythological, symbolic, and iconographic representations to earlier civilizations. The interrelatedness of the cultural and natural environment is a primary characteristic of pre-Columbian religions and cosmology (Moore 1984; Sullivan 1985, 1988; Lane 2002; Foster 2005; Staller 2008a, 2008b). Pre-Columbian lightning veneration and its associated cults and rituals represent a worship of nature and the forces that shape the natural world (Frazer 1926; Zuidema 1980, 2011; Taube 1996b; Sullivan 1985, 1987, 1988, 2002; Conrad and Demarest 1984, 1992; Markman and Markman 1989). Our analytical approach was undertaken in part to determine whether similarities and differences in the patterns and associations brought out by our comparative framework would address larger questions regarding distinctions and similarities in pre-Columbian religious thought, and how legend and folklore have been transformed to varying degrees by religious conversion and acculturation in the post-Conquest and modern world (Brinton 1894; Krappe 1936; Foster 1945; Toor 1947; Costas Arquedas 1967; Gade 1983).

The striking linguistic diversity of Mesoamerica and the relative linguistic homogeneity of the Andes made direct comparative analysis of pre-Columbian religious beliefs, cosmology, and indigenous worldviews problematic from a contextual standpoint. There are three or four major families in Mesoamerica and over seventy indigenous languages. On the other hand, in the Andes Quechua and Aymara are the most widely spoken language families, and they share similarities revealing long periods of interaction. There are many fewer

[5] The evidence presented in this study indicates that lightning has strong symbolic and metaphorical associations to metallurgy and the metallurgical arts in the Andes.

languages overall in the Andean region. This disparity was revealed in part when we considered the relative linguistic homogeneity of the Andes compared to the heterogeneity of Mesoamerica and the extent to which nearly ubiquitous similarities in lightning veneration in the Andes and its associated rites are related to such linguistic differences.

In Mesoamerica, the contextual variability of the meanings and associations is considerably greater in the particulars, with multiple languages represented in multiple language families, thus existing in many more separate and for the most part closed corporate communities. In stark contrast to Mesoamerica, in the central and southern Andes little remains to us of the languages other than Quechua dialects and Aymara that must have once existed there. Such linguistic homogeneity is reflected in the similarities of language, rites, and rituals surrounding lightning over a vast geographic area of the South American continent. The linguistic diversity in Mesoamerica and continuous widespread language convergence of Aymara and Quechua in the Andes is also related to some extent to cultural and economic as well as geographic and environmental factors that facilitate the integration or the isolation of language groups. Linguistic and cultural differences make direct comparisons of the significance of lightning to pre-Columbian cultures in the Andes and Mesoamerica challenging, but most important is the fact that describing cultural traditions is best done within relevant cultural contexts prior to comparing culture elements shorn of some of their context, relying on the language of comparison (here English) to supply a kind of "etic" context.[6]

This study documents analytical and symbolic relationships through an emphasis on how and why cultural beliefs concerning lightning have particular symbolic and metaphorical associations to pre-Columbian mythology and cosmology. It became apparent early on that direct comparative analysis would pose contextualization problems related in part to the linguistic differences that characterize the two culture areas. Thus, we have deliberately organized the volume as separate presentations of lightning in these culture areas before making comparisons. Our goal was to provide contextualized descriptions of lightning ideology and veneration that would not be reduced to listings of present or absent traits from each region to be directly compared. Identifying and separating "traits" from their respective cultural contexts could potentially misrepresent their cultural and religious meanings. Furthermore, our respective descriptions are colored by our individual understandings of meaning, relevance, and coherence of the available evidence. Direct comparison could potentially obscure information derived from our individual understandings.

---

[6] Etic description is from the point of view of an outside observer of a culture, with a view toward comparisons with other cultures, attempting to be as neutral and objective as possible. On the other hand, an 'emic' description, is one that is given in terms of a culture's language, is meaningful to members of the society itself, and shares values and evaluations made in that society.

Therefore, we felt it necessary to clarify the meanings related to our individual interpretations from our respective analyses of each culture area by presenting them separately and in context. Within a framework of separate accounts of these two culture areas we have then considered elements taken from each account, which can be contrasted and compared, having already been presented in their respective contexts. The various similarities and differences as well as the broader implications of the data generated by this comparative analysis are addressed at the end of each section and synthesized in the discussion and conclusions. Our presentation of these data and the organization of the book are consistent with traditional anthropological processual approaches to comparative analysis (Binford 1962, 1964, 1965; Coe 1994; Cook 1998; Conrad and Demarest 1992).

Our study of lightning in the Andes and Mesoamerica includes sections that explore places in the natural world, such as caves, mountains, and bodies of water, as well as locations that were modified by indigenous populations, who in the Andes perceived such places to be endowed with a spiritual life force or the animating essence associated with lightning. Landscapes provide meanings through a connection with creation and origin myths, folklore, legends, and rituals performed to venerate and animate the specialized spiritual forces or powers such as lightning, keeping their symbolic and metaphorical associations alive, and at the same time transforming them through syncretism as cultural histories and identities changed. Caves and water sources, so strongly connected to lightning and the rain deities in Mesoamerica, also have significance to lightning in the Andes, primarily as a metaphorical reference to emergence into a world or creation cycle or to accessing the tripartite cosmos, particularly through feline and reptilian animal familiars. Lightning is also commonly associated with meteorological phenomena such as life-giving rain, rainbows, and bodies of water, as well as destructive elements such as hail, fire, and resultant crop failure and destruction of the natural environment. Lightning and its various symbolic and anthropomorphic manifestations were of central importance to agriculturally based civilizations such as those under consideration herein, primarily because agriculture depends on rain and lightning is a harbinger of rain, but, perhaps more importantly, lightning's power to create as well as to destroy, to aid a community or to damage it.

We found that lightning was associated with pre-Columbian rulership in both regions and, at least in recent times, with shamanism. Lightning bolts in both regions, as well as in several other parts of the world quite independently, are seen as a source of stone tools, including obsidian, flint, and other silica-based stones and transparent or translucent quartz crystals. In Mesoamerica lightning is commonly associated with riches and good fortune. In Quechua regions of the Andes, riches and good fortune are linguistically related to the root for the current term for lightning; precious metals and sites associated with the metallurgical arts are associated with lightning in certain regions of the Andes. Child sacrifice was practiced in connection with lightning in both regions. Contemporary shamans in both culture areas are frequently "recruited" by surviving direct lightning strikes. Lightning is closely linked in contemporary cultures in these regions to the ideology of consumption associated with ecstasy and ecstatic trance, which can be induced by lightning.

This introduction only begins to approximate the scope and content of the volume. It was one of our goals to demonstrate the importance of lightning and the coherence of associated concepts in parts of the Western hemisphere in order to provide a basis for further discussions of the place of lightning as embedded in other cultural traditions. We also wanted to illustrate an approach to regional comparison that stresses separate presentations of material from each region with a view to presenting information in context as well as providing acknowledgment of author predispositions along with transparency of analytical presuppositions.

The Nature of Lightning

Lightning as we understand it today is a discharge of static electricity, not so different in character from the electricity that shocks your fingers when after walking on a carpet you reach for the doorknob and feel it, hear it, and maybe even see the spark. But the difference in magnitude and the power of the shock produced by lightning is obviously enormous. The giant discharge of static electricity that we know as lightning is produced within clouds and also occurs between clouds and other clouds, and more spectacularly between clouds and the earth. Possibly related to the differences in electrical potential created by friction of tiny ice crystals within clouds, lightning and the thunder that follows it usually attend rainstorms, or thunderstorms.

Thunderstorms are a result of either the displacement of warmer air by an invading cold front, or, more commonly in the tropics, strong convection currents intensified by intense solar radiation. While lightning itself is not fully understood, the word as used here assumes the following: lightning is an electrical phenomenon of great power and heat energy, a discharge of static electricity usually associated with clouds and with rain. In the tropics, thunderstorms frequently generate anvil-shaped cumulonimbus clouds, unleashing powerful updrafts and fearsome strokes of thunder and lightning. The convection currents caused by the displacement of warm air by a cold front creates a friction, generating electrical charges or electrons. Rain or hail carry charged electrons to the lower portions of clouds, where the negative charge builds, while positive charges build on the tops, and lightning bounces back and forth within the clouds, a phenomenon generally referred to as cloud-to-cloud lightning. When the flash of such lightning is hidden in the clouds, the broadly visible illumination is often called sheet lightning. Most of the electrical energy in a thunderstorm is dissipated in this manner within and among the rain clouds. When the negative charges in the lower portion of the clouds are great enough, they become attracted to positive charges emanating from the surface of the earth, and the effect is a lightning bolt.[7] Bolts are first and foremost attracted to hilltops or mountain summits, and, in the case of tropical rain forests, the tallest trees or architecture—that is, the highest points on the surface of a particular landmass.

The visual expression of lightning is generally accompanied or shortly followed by sound waves, generally assumed to be from the rapid expansion of the air that has been immediately displaced by the lightning. This heated expansion of air around lightning and lightning bolts is superheated to about 54,000°F (29,982°C), expanding faster than the speed of sound,[8] and the result of the sudden heating and expansion is heard as thunder, the audible pressure (compression) wave causing the sonic expression.[9] When a discharge of lightning or lightning strike hits the earth, and even sometimes when it remains in the clouds, it is often described as a lightning bolt (emphasizing the visual expression) or as a thunderbolt (emphasizing the sonic expression) (**Figure** 4).

Almost all lightning is generated by thunderstorms. However, lightning, both cloud-to-cloud and bolts hitting the earth, have been observed during snowstorms and in smoke from forest fires, as well as in and around clouds of volcanic ash created by volcanic eruptions. Lightning also results in ionization in the atmosphere, leading to the formation of nitric oxide (NO) or nitrogen monoxide, and ultimately to nitrogen dioxide (NO_2) when exposed to oxygen (O_2). Another chemical pathway of NO is to the formation of nitric acid (HNO_3), which is beneficial to plant life in the form of nitrates in fertilizer (Zumdahl 2009; Bond et al. 2002). Ultimately all life depends on the availability to plants of nitrogen in "fixed" form. Since the nitrogen molecule is quite inert, large amounts of energy input are needed to break it apart so that its atoms can combine with atoms of other elements, notably oxygen and hydrogen. Lightning can supply the energy to thus "fix" nitrogen, making it useful for plants, and paradoxically can also create pollutants (Bond et al. 2001, 2002).

The increased intensity and association of lightning storms with forest fires has been found to be related to reactive nitrogen oxides (NO_x) emitted

---

[7] Although all lightning bolt discharges effectively transport cloud charge to the ground, 90 percent of them are of a single type, known as downward negative lightning. Upward lightning strokes are believed to occur only when delivered from very tall objects, i.e., three hundred or more feet high (Rakov and Uman 2007:4–5).

[8] This is five times hotter than the sun, and the light thus produced travels 186,291 miles per second (299,806 km/sec); sound travels only 1,088 feet per second (332 m/sec), thus the perception of lightning usually precedes thunder (Bond et al. 2002).

[9] The differences in speed is why thunder is sometimes heard to rumble, in that each shock wave takes a different amount of time to reach the listener.

FIGURE 4 Cloud-to-cloud lightning is the most common and referred to by native Andeans by the Spanish term *relámpago*.

into the atmosphere from natural and anthropogenic sources (Seinfield and Pandis 1998; Bond et al. 2002). Biomass burning related to slash-and-burn agriculture is a common source of reactive nitrogen oxide outside major population centers (Bond et al. 2002). Reactive nitrogen oxides are also released by industrial pollutants, discharges of lightning bolts, and cloud-to-cloud lightning, as well as released through evapotranspiration by the soil.

Research results indicate the possibility that convection in thunderstorms can transport NO_x-rich air as a result of anthropogenic, biomass burning, or soil emissions from the planetary boundary layer into the middle and upper troposphere. This leads to an enhancement of NO_x in the free troposphere that appears to have a profound impact on ozone (Bond et al. 2002:1516; Tie et al. 2001; see also Dickerson et al. 1987). This impact is related to an overall thinning of the upper ozone layer. However, the magnitude of those enhancements is dependent on the amount of pollutants in the planetary boundary layer and the dilution rate in transporting of NO_x to the middle and upper troposphere, which is difficult to quantify (Bond et al. 2002:1516). Biomass burning emits the greatest annual amount of NO_x and lightning discharges rank third in magnitude among the four variables considered in various studies. However, lightning occurs throughout the annual cycle and is the most significant source of NO_x over the world's oceans (Bond et al. 2002:1517).

Recent technologies have provided increasingly detailed information regarding the occurrence and intensity of lightning storms in different parts of the world. In past centuries, sailors had reported seeing the most intense and frequent occurrence of lightning in the tropical latitudes (Orville 1990). Recent quantitative data generated from satellites, Optical Transient Detectors, and Lightning Imaging Sensors have reported significant detailed information on a global scale along these lines (Pinto 2009a, 2009b; Pinto et al. 2006). Lightning satellite observations have indicated about forty-five ground and cloud flashes occur throughout the world every second, more than a billion flashes per year, and about 70 percent of these are in the tropical latitudes. The more technologically sophisticated sensors validate sailors' observations, indicating that 76–85 percent of lightning worldwide occurs in the tropics (Christian and Latham 1998; Christian et al. 1999). Thus, NO_x production by lightning occurs predominantly in the tropics. For example, in the Western hemisphere, lightning produces at least 40 percent of the NO_x over much of Central and South America (Bond et al. 2002:1513). Observations for the various landmasses along the tropics indicate that the largest total lightning activity in the Americas occurs in Colombia and Venezuela, that is, the northern Andes (Pinto et al. 2006, 2009a, 2009b). It has also been found that lightning activity in Brazil is of a longer duration than in either Colombia or Venezuela (Pinto 2009b:173–174). Scientific research on reactive nitrogen oxides NO_x suggests that this may be related in part to the deforestation of the tropical forests in Amazon region of lowland South America. Production of NO_x by tropical lightning is significant throughout the year. Although lightning strikes are far more frequent over land than at sea, lightning accounts for almost all of the NO_x emitted over the oceans and 50–90 percent emitted over continental landmass areas on a seasonal basis (Bond et al. 2002). Annually the contribution of lightning to total NO_x production is 23 percent in the tropics, and overall NO_x production by lightning occurs predominantly in the tropics.

There is good reason for humans to fear lightning and to worry about being struck and perhaps killed by it, especially if one is outdoors and the tallest thing around during a lightning storm. Lightning can potentially affect all the organs, particularly in the cardiovascular system. The primary cause of death following a lightning strike is cardiopulmonary arrest. Changes in the heart rhythm and even heart stoppage may temporarily occur, but usually normal rhythm will quickly be resumed. Paralysis of the respiratory system appears to be more common and longer-lasting than heart stoppage. If artificial respiration is not initiated immediately, a person will usually die of hypoxia or lack of oxygen (Charleton 1995: 265–266). Another common effect is first- and second-degree burns on the skin or scalp. Bleeding within the brain is another possible result of a lightning strike (Rakov and Oman 2003) Approximately 50 percent of all surviving lightning victims have some kind of damage to their eyes, usually on the cornea. The most common serious eye injuries are cataracts, which may last for only a few days but can last as long as several years after the lightning strike (Browne and Gaasch 1992; Graber et al. 1996; Rakov and Oman 2003).

Worldwide, there are some 16 million lightning storms annually. More deaths are caused by lightning strikes than by any other natural phenomenon, including tornadoes, hurricanes, and floods (Rakov and Oman 2003:642–649). Approximately one thousand to fifteen hundred people are seriously injured by lightning strikes annually in the United States, and of those, between one hundred to six hundred people perish (Browne and Gaasch 1992; Rakov and Oman 2003). Thus mortality rates range between 25 to 32 percent. Approximately 75 percent of survivors of a direct or indirect strike sustain permanent injuries (Browne and Gaasch 1992:225; Rakov and Oman 2003:449; Taussig 1968).

Men are said to be about four times more likely to be struck by lightning than women (Friedman 2008:6), but then men are also more likely to spend time outdoors. Rakov and Oman (2003:2) report that 103 church bell ringers were killed in Germany by lightning bolts that struck 386 church towers during a thirty-three-year period in the late eighteenth century (cf. Friedman 2008:43). Such statistics make one wonder not only what incentives could attract one to that profession, but also why church bells would be rung during thunderstorms. In answer to why they would be rung, for many hundreds of years Europeans believed that ringing church bells could ward off lightning, and this effect could be augmented by prayers of the faithful on hearing the bells (Friedman 2008:42).[10]

Although in some contexts it is perfectly understandable for one to fear lightning, there are situations in which it is not such a rational fear, such as when riding in a subway. An irrational fear of thunder and lightning is referred to as astraphobia, and it has been known to affect both humans and their domesticated animals, particularly dogs and cats. Between the rational and irrational fear of lightning, it is clear that lightning is always an object of potential terror, for animals as well as humans. In addition to killing people directly, lightning can also start fires with the same effect. Humans, finding it useful for cooking food, preparing land for planting, and keeping wild animals at bay, have a healthy respect for fire but generally aren't alarmed unless it gets out of control. Lightning, on the other hand, is always out of human control.

Flint striking iron produces bright flashing sparks that can start fires, and lightning has the visual appearance of giant bright traveling sparks such as might be struck from iron or steel by a gargantuan piece of flint. It is not surprising, then, that lightning is likened by some to giant sparks and thought to be a similar power contained within the flint itself. The analogy is inaccurate

[10] In many and diverse parts of the world there are peoples who practice loud noisemaking to ward off various perceived dangers, such as eclipses, lightning strikes, wild animals, evil spirits, etc. In China, Borneo, and parts of Mesoamerica, for example, loud noises, including beating of pots and pans together, are made as an eclipse commences, for it is believed that a monster is eating the moon. Until very recently, Bantu priests made loud noises, shouting and playing flutes, to try to scare off the thunderbirds responsible for lightning.

according to current science, however, in that lightning is primarily electrical in nature, whereas sparks struck from iron, a material that spontaneously ignites below room temperature, are seen as due to rapid oxidation of minute iron particles, a process facilitated by the small size and therefore large surface area of the particles. Banging pieces of flint or quartz together will also produce sparks, but not hot enough or long-lasting enough to start a fire. Sparks, both hot and long-lasting, are produced in metal smelting processes, and since metallurgy was practiced on a relatively large scale in the Andes, it is unsurprising to find that metallurgy is, and was in the past, connected to lightning.

Lightning and thunder are almost invariably accompanied by rain, though rain may well occur without accompanying lightning. When there is lightning, the weather phenomenon is often termed a thunderstorm, a lightning storm, or a rainstorm. Nevertheless, electrical discharges or lightning strikes rarely occur in a clear blue sky. Statistics indicate that the majority of people struck by lightning bolts are struck just before or after a thunderstorm, not during the heaviest rain. Many languages make distinctions among "lightning," "thunder," "lightning bolt," "thunderbolt," "rain," and "rainstorm" that are similar to our own, though not necessarily the same, while also recognizing, as we do, some ambiguity in application of the words.

2

Andean South America

Lightning in the Ancient World

The celestial realm is symbolized by two images of uniqueness and indivisibility: liquid and light. Principal manifestations of heavenly light include sound forms and visions of light (Sullivan 1988:115–116). The significance of lightning in Andean religious thought is that it represents an absolute manifestation of the heavens and thus displays and evokes the meaning of *eternity*, "a state that suppresses differences in that even pairs of opposites coincide" (Sullivan 1988:32). These qualities and characteristics explain in part why lightning was such an important agent of transformation and powerful entity in the ancient world and why it has associations to creation and world cycles in various mythologies. Heavenly fluids such as the celestial river, the Milky Way (Mayu), and dark cloud constellations provide symbolic connections between the celestial (*janan pacha*) and the underworld (*uchu pacha*) realm. Among Andean societies, dark cloud constellations are called Pachatierra and seen as an ambivalent intermediate female manifestation (fluidity) within the realm of structure. Light, the primary visual characteristic of lightning, is associated among Quechua and Aymara speakers with structure, the channeling of fluids, and the male generating principle (Nuñez del Prado 1974; Urton 1981, 1988, 1992; Classen 1993; Staller 2008b). The Milky Way is seen as a celestial river that moves water from the earth to the underworld and through the sky (Urton 1988:93). Perhaps it is the essential characteristics of thunder and lightning and the degree to which bolts affect human lives and consciousness that make it such an important religious icon (**Figure** 5; see color plate section). The ways in which lightning veneration is expressed and the ritual patterns and associations are carried out are to varying degrees distinct. However, they also have much in common, particularly regarding folklore and legend as well as their associated symbols and their relationships to contemporary indigenous culture and ritual practice (Gade 1983; Topic 1992). The transcendental qualities of lightning are later manifest in a syncretism of folk practices and religious beliefs whose high degree of cultural integration has emerged from a veneration of the natural world (Gade 1983:771; Sullivan 1987).

Lightning received such widespread religious veneration and representation in the ancient world because of its essential qualities and natural association to the heavens, and also to the underworld through lightning bolts that penetrate bodies of water. During Inca expansion and early Colonial period, lightning veneration referred to Illapa among the Inca and was related to the Catequil cult throughout their empire (Topic et al. 2002). The various symbolic associations associated with this indigenous cult are also apparent in the related symbolism and iconography surrounding rainbows, felines, serpents, underground springs, and caves among contemporary Andean societies. There are intrinsic relationships between light, color, and water. Because these phenomena are all regarded as manifestations of celestial forces due to their emergence from the earth or sky or water, they establish an interconnection among the various levels of the universe. The Andean universe was generally perceived as a tripartite cosmos with the earth in the center as a mirror that alternately reflects the celestial and underworld order across and through itself (Urton 1981:63, 93; Lechtman 1999:227; Staller 2008b:287). It is perhaps in part from the Native American worldview that astronomical calculations of the heavens and of lunar and solar cycles played such an important role in cultural perceptions regarding the interrelatedness of the cultural and natural world—particularly among cultures with agricultural economies.

TABLE 1

Andean Lightning Associations

1	Lightning in the Ancient World
2	Inca Cosmogony and Ethical Order
3	Language and Lightning
4	Lightning in Pre-Hispanic Religion
5	"Sons of Lightning" and Immaculate Conceptions
6	Shells of Lightning: Fertility, Rebirth, Death
7	Lightning *Huacas* and Inca Expansion
8	Lightning, Metal, and Death
9	Colonial and Contemporary Religion
10	Lightning in Inca Cosmology and Mythology
11	Earth, Sky, and Water in Andean Cosmology
12	Temporality of Lightning
13	Lightning and the World Inside
14	Lightning Shaman: Human and Animal Familiars
15	Fictive Kinship in the Cultural and Natural World
16	Pachatira: Rainbows, Serpents, and Water
17	Lightning Stones

Inca Cosmogony and Ethical Order: Lightning, Creations, and Chaos

Pre-Hispanic religions of the New World were inherently telluric, that is, naturalistic and spatial (Sullivan 1984, 1985). Their origins and creations are from previous conditions or states, rather than out of nothing (*creatio ex nihilo*) or first beginnings (Staller 2008a:7; Fabian 1983; Sullivan 1985, 1988). The Inca developed a profound ontology of periodicities and spatialities. These were revealed to chroniclers and clerics through rituals and ritual performance—an ontology of what was "true" and "real" in celestial and terrestrial space and the "natural" order of social relations that they transformed into major principles of ethical order (Sullivan 1985:99). The chroniclers to varying degrees mention pan-Andean conceptions of creation and emergence; however, Inca cosmogony is what is best recorded from the chronicles and sixteenth-century accounts (Betanzos 1987 [1551]; Cieza de León 1998, [1553], 2005 [1553]; Garcilaso de la Vega 1963 [1609]) and then later to Spanish colonial governments and conversion to Catholicism (see, e.g., Sarmiento 1907 [1572]; Molina 1959 [1575]; Anonymous 1906 [1580–1621]; Albornoz 1989 [1584]; Acosta 1962 [1589]). The earliest chroniclers who discuss Inca cosmogony focused on perceptions and beliefs in the imperial capital, Cuzco, the spiritual and geographic center of their empire, Tawantinsuyu, the "Land of the Four Corners" (**Figure** 6). The Inca state was divided into four parts: Chinchasuyu and Antisuyu, associated with *hanan* or "upper," and Collasuyu and Chinchasuyu, associated with *hurin* or "lower" (Magli 2005:23). The Inca Empire flourished between the fourteenth century CE and the Spanish Conquest in 1532. Each of the empire's four quarters was divided into forty-one radial sectors, and between the sectors the sight lines were called *ceques* (Zuidema 1964, 1995, 1997, 2011; Bauer 1998). The Cuzco calendar was codified by the organization of 328 *huacas* along these sight lines or *ceques* from the center of the empire, the Haucaypata or "plaza of leisure," whose maintenance was not associated with *mitá* obligations (González Holguín 1989; Zuidema 2011). The organization of the Inca Empire into upper and lower extends to the main plaza in the imperial capital of Cuzco, Haucaypata, which is therefore of critical importance to understanding the organization of the capital as well as the empire (Zuidema 2011). The principles of ethical order that organized the city of Cuzco and the empire itself were bound to the experience and organization of celestial and terrestrial (natural landscape) spatialities and temporal relationships manifest in the creation of the cosmos (Sullivan 1985:100). Early chroniclers in Cuzco gathered most of the information on Inca creation myths and the creation myths of their subject highland and coastal populations, and there is clear evidence that some aspects recorded by the chroniclers on the Inca cosmogony were pan-Andean (Valcarcel 1964:366–369; Pease 1973:18; Sullivan 1985:100).

Indigenous accounts and later oral traditions recorded by missionaries, historians, anthropologists, and Andeans indicate a widespread concept of

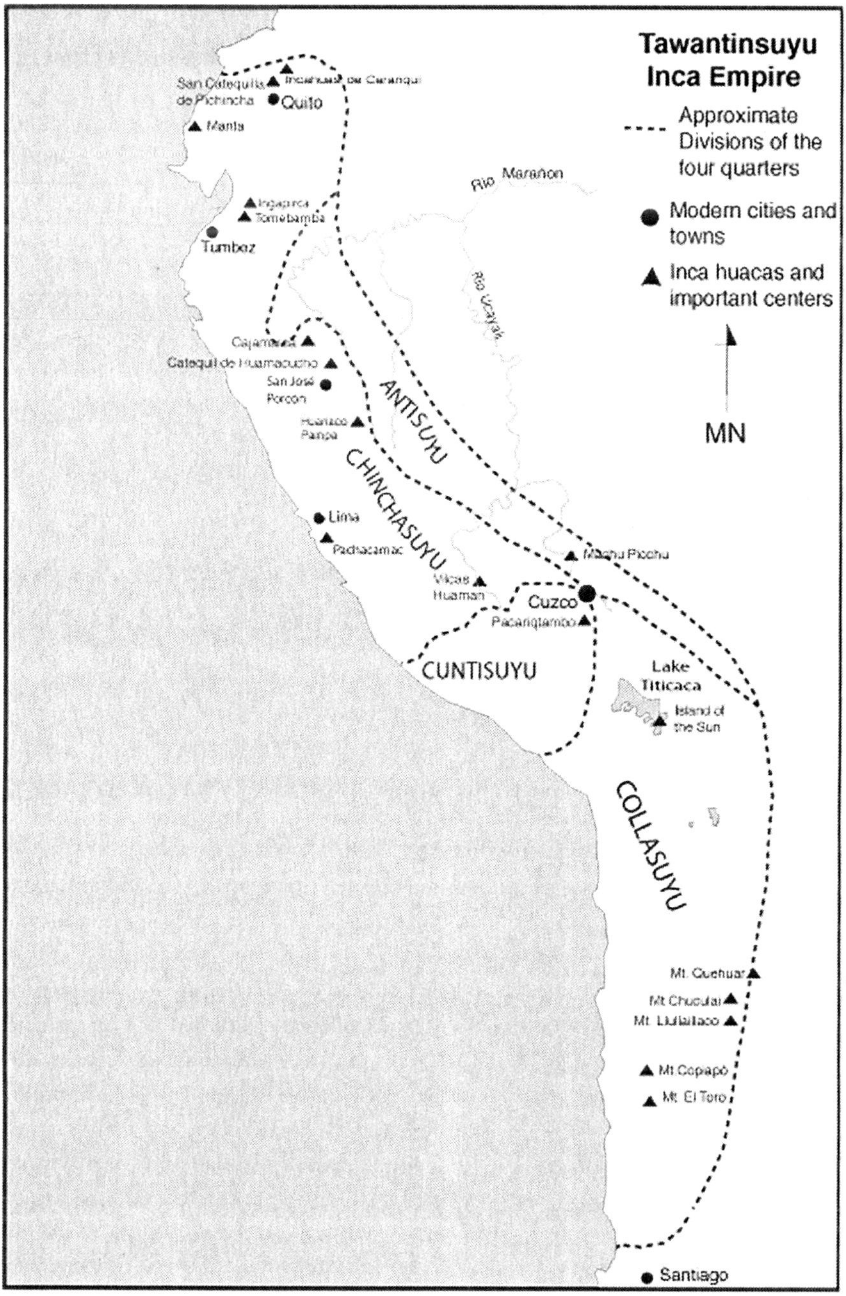

FIGURE 6 Tawantinsuyu, or "Land of the Four Corners," showing the approximate locations of the four corners of the Inca Empire as well as important *huacas* and *ushnus* mentioned in this study. The Inca state was divided into four parts: Chinchasuyu and Antisuyu, associated with *hanan* or "upper," and Collasuyu and Chinchasuyu, associated with *hurin* or "lower," and these divisions had important implications for the organization and administration of non-Inca populations, particularly regarding tribute and redistribution.

four creations or world cycles, origins, and creations from previous conditions or states of being, separated by three periods of destruction or chaos. Beyond widely held beliefs regarding four ages or cycles of creation, there are also concepts of a mountain body and a creator, Viracocha, with dialectic properties, whose name has reference to seafoam and at the same time to the celestial realm (*janan pacha*). Joseph Campbell (1949:146) mentions that Viracocha is also venerated as both a solar and storm god among some cultures (see also Demarest 1981:35–39). Bolts are associated through mythologies of emergence to the "living waters," which represent the "tears of god." The lightning bolt at once annihilates and is itself indestructible: the delusion-shattering light of the Imperishable is synonymous with the light that creates, thus fire and water are one and the same (Campbell 1949:146). Among some Aymara cultures Viracocha is also associated with thunder (Demarest 1981:35).

Andean scholars have maintained that pan-Andean cosmogonic themes included a perception of an emergence into a wild (*purum*) landscape or natural state of being into the existing world cycle or creation (Sullivan 1985, 1988). Such mythological emergence in Andean religious thought are usually from places such as mountain summits, hilltops, mountain passes, rock outcrops, caves, lakes, lagoons, and natural springs on what is conceptualized as an Earth Mother (Pachamama) (Sullivan 1985:100; Urton 1990, 1992, 1999; Salomon and Urioste 1991). Religious veneration among the Inca and other Andean cultures was truly a worship of nature and the natural world (Frazer 1926, 2000; Sullivan 1987). Chroniclers report that Inca rulers had fictive kinship relations with sacred places (*huacas*) and that such places were sometimes seen to have such fictive kin relations to one another. Rather than venerating a pantheon of anthropomorphic deities, the Inca and other Andean cultures believed that an animating essence (*enqa* or *sami*) permeated all things (Flores Ochoa 1978). Thus all life, as well as caves, mountains, lagoons, and even living rock, was to varying degrees sacred. The Inca worshipped the sun, moon, and meteorological phenomena such as lightning and rainbows, as well as certain places in the natural landscape, which linked their ethnic identity to the surrounding landscape. Most objects of veneration—lightning, sun, moon, and so on—were bestowed with gender associations (male/female), as were mountains, caves, valleys, and certain sacred places in the natural landscape. There is archaeological evidence to suggest that such religious perceptions and beliefs are very ancient in this region of South America. Veneration of the natural world perceives a sacred landscape "fed" through ritual offerings and transformed or civilized through human action (Sullivan 1985:103–109; Pease 1973; Classen 1993; Pachacuti Yamqui 1950 [1613]). Ritual performance and offerings impose a moral order of being intrinsic to the ethical order expressed by Inca ontology and cosmogony.

The Andean landscape symbolically and metaphorically refers to the human body (*ucu*) and the concept of a mountain body; mountain summits represent *uma pacha*, the head or original time and place; highland springs (*puqyos*) represent the eyes; and caves (*pacarinas*) symbolize the mouth (Sullivan 1985:106). Sacred places (*huacas*) in the landscape are thus "fed" through specific kinds of ritual performance and offerings. Andean communities feed or make ritual offerings to *huacas* in order to perpetuate the circulation of time and bring forth regeneration and the fertility and fecundity of the earth (Bastien 1973:170; Sullivan 1985:107). As the following study indicates, lightning has close symbolic associations to all such places of emergence in the landscape and more generally to the head or *uma pacha* (original time and place) of the sacred landscape (Classen 1993; Pachacuti Yamqui 1950 [1613]).

The fourth creation represents the existing world cycle, marked by dynamism and the transformation of the natural world through human action, which is generally based on traditional forms of tribute and reciprocity (Covey 2006). The four world cycles are separated by three periods of chaos, generally perceived as divinely imposed destructions of unbroken duration, involving conversion into stone (earthquakes?),[1] then water (deluge and floods), and, finally, fire (Pease 1973:18; Sullivan 1985:103; Pachacuti Yamqui 1950 [1613]:209–211). Inca cosmogony suggests lightning may also have reference to intermittent cycles of chaos and destruction that separate the various world cycles: stone is generally perceived in Andean culture as the oldest living matter on earth. Rock outcrops cleft by lightning were enclosed architecturally through platforms and enclosures of cut stone to mark their sanctity. Lightning bolts that strike water, ponds lagoons, lakes, and so on link the celestial to the underworld and are associated with earthquakes through Catequil de Huamachuco, the principal lightning *huaca* in the Andes, which was perceived as a *yañca*—that is, a *huaca* believed to make other *huacas* speak. Lightning is associated to fire by bolts, which cause forest fires or burn crops. In Inca cosmogony, lightning was descended from Viracocha the creator, and Inti the sun; it was also brother to the Earth Mother, Pachamama (Sullivan 1985, 1988; Urton 1990, 1999). Lightning has clear symbolic reference to intermittent cycles of chaos and destruction that separate the various Inca world cycles. Thus, lightning is closely associated symbolically and metaphorically with earthquakes, water, and fire and descended from Viracocha, symbolic relationships clearly brought out in Inca cosmogony and cosmology (Sullivan 1985; Classen 1993; Salomon and Urioste 1991). These associations also imply that lightning was perceived to have earlier origins, explaining why in some regions of the Andes lightning *huacas* and *ushnus* had earlier pre-Hispanic occupations and architectural modifications that preceded Inca expansion into those regions.

[1] It is for this reason that origin mythologies often involve the original mythological ancestors as transforming into stone, thus connecting their people to earlier—and, in the case of the Inca, the earliest—creations and destructions (Urton 1990, 1999; Salomon and Urioste 1991; Classen 1993).

Andean cosmogony perceives emergence from this wild or natural state into the existing world creation cycle ultimately to impose and/or create, *ayni* (i.e., balance and reciprocity). In Andean culture, the creation of *ayni* is generally through a culturally imposed ethical order, a "civilizing" or the systematic creation of civilization (*camachisqa*) over people (*camachicuna*) and lands (*patachana*), through the cultivation of valley bottomlands, cultural modifications of sacred places, channeling water through irrigation canals, and construction of artificial terraces in steep valleys of the cordillera (Classen 1993:11; Salomon and Urioste 1991:43–44; Covey 2006) (**Figure** 7). In 1551, the chronicler Juan de Betanzos[2] described Inca perceptions of creation cycles and intermediate periods of divine imposed destructions of unbroken duration. Betanzos's descriptions were summarized by Valcarcel (1964, 2:367–369) as consisting of total darkness, and it was during this, the third creation, that "Con Titi Viracocha (the creator) came forth from a lake." This description is compelling in that it relates to the legends and myths of emergence surrounding El Dorado, the golden one, who emerged from a cave covered in gold dust and then jumped into Lake Guatavita in a rite of passage, as well as legends of "lost cities of gold" from various regions of the Andes and other parts of South America (Anderson and Bray 2006). Such associations have direct relevance to lightning because the idol in Cuzco was a faceless gold statue, and, as documented in this study, many lightning *huacas* were associated with the metallurgical arts (Bouysse-Cassagne 1993; Cruz 2009). The various legends of El Dorado underwent numerous transformations, and some Spanish accounts of the legend were recombined with the legendary "cities of gold" that so fascinated and were so often mentioned by Spanish, Portuguese, and later English explorers for over two centuries. Other Andean origin myths also speak of emergence into a wild landscape from lakes, ponds, and lagoons and in a number of pre-Hispanic cultures have direct associations to creator gods. Lightning has close associations to such places through bolts that descend from the sky and strike bodies of water, thus linking the celestial realm to the underworld (Casaverde Rojas 1970:150; Nuñez del Prado 1974:246–247; see also Urton 1981). Valcarcel (1964:368) goes on to say that when Viracocha arrived at Cacha (eighteen leagues from Cuzco), he "made fire rain down from the sky and a ridge of mountains was burned." On seeing this the Cancha threw themselves on the ground; then Viracocha "took a staff in his hands to where the fire was and struck it two or three times," extinguishing it (Sullivan 1985:102).

---

[2] Juan de Betanzos was married to a former wife (*mamacona* or *acylla*) of the Inca emperor Atahualpa, spoke Quechua, and lived in Cuzco among the colonial Inca elite. The chronicler Garcilaso de la Vega also spoke Quechua, was educated in Cuzco, and was married to an Inca princess, baptized as Isabella in 1539 (Cieza de León 1998 [1553]:159 n.). Isabella was the daughter of Hualpa Tupac, younger brother of the last Inca ruler, Huayna Capac. Both Betanzos and Gacilaso de la Vega were sympathetic to indigenous culture and thus provide less biased and more informed insights into Inca customs and rites (see, e.g., Staller 2008b, 2010a–b).

FIGURE 7 Colca Canyon located northwest of Arequipa in southern Peru. Some terraces in this valley have pre-Inca origins associated with Wari expansion. The Inca cultivated maize and other crops on such artificial terraces along the 4,160-meter-deep sides of the canyon. The pre-Inca origins of such modifications imply earlier Wari expansion, which also involved "civilizing" the landscape.

In later Inca cosmology it was Illapa who made fire rain from the sky and who would later extinguish it with the coming rains. Cultural beliefs of four epochs or creations appears to continue after the Conquest, as is evident in the Huarochiri manuscripts as well as the writings of Guaman Poma de Ayala and Santa Cruz Pachacuti (see Guaman Poma 1980 [1583–1615]; Pachacuti Yamqui 1950 [c. 1613]; Salomon and Urioste 1990 [1598–1608]). Pan-Andean cosmogonies have in common the perception that the first three creations were not only of temporally unbroken duration, but also spatially totalitarian in that they are governed by one eternal principle of being, the earliest stone, then absolute light and total darkness that ruled unbounded and unchallenged in space (Sullivan 1985:104). It is only with the fourth creation that different human cultures appear, and terrestrial space becomes fragmented into Hanan/Hurin ("lower/upper"), Antisuyu, Contisuyu, and Chinchasuyu and Collasuyu (see **Figure** 6). It is only in the current human creation cycle that time becomes fragmented into periodicities and governed by the spatial dispositions of the sun, moon, and stars, and their manifestation as lightning, clouds, hail, rain, drought, and day and night appear (Sullivan 1985:104). Human souls appear from *uma pacha* (original time and place) places associated with mountain summits, mountain passes, caves, and highland springs—the landscape features with the strongest associations to lightning. Inca cosmogony provides compelling evidence that lightning (Illapa) has closer symbolic associations to Viracocha than the sun (Inti) and moon (Quilla), and thus to earlier creations and particularly to intermediate periods of destruction and chaos.

Language and Lightning in the Andes

Any attempt at understanding the variability and similarity of lightning veneration and its associated religious rituals and rites in the ancient Andes requires some consideration of the surrounding language and linguistics. It became apparent early on that one of the primary challenges in dealing with a comparative study of any aspect of pre-Columbian religious beliefs, cosmology, and/or indigenous worldviews in these culture areas is the striking linguistic diversity of Mesoamerica and the relative linguistic homogeneity of the Andes. In Mesoamerica there are over seventy indigenous languages divided into three or four major families, but in the Andes there are many fewer.

Quechua and Aymara are most widely spoken throughout this part of South America (Torero 1974:29–34; Cerrón Palomino 1987, 2000a, 2000b, 2003). Andean language families show considerably less time depth than those in Mesoamerica. However, Cerrón-Palomino (2000b:337) has identified clear and linguistically deep associations between the Quechua and Aymara language families reflecting an intense contact from the earliest stages of their dispersal(s). Quechua and Aymara have a great deal of vocabulary in common, particularly borrowed terminology and phrases, as well as a host of striking structural parallels (Cerrón-Palomino 2003:311, 2008). The degree of convergence reflects long-term contact between what were originally separate populations. However, as Proto-Quechua and Proto-Aymara spread geographically they also went through periods of divergence, reflecting independent expansions of two originally different population groups of different origins—Quechua speakers on the one hand, and Aymara speakers on the other (Cerrón-Palomino 2000a:140, 2000b:337; Heggarty 2007; Beresford-Jones and Heggarty 2011).

Some linguists have interpreted expansion of Quechua with Early Horizon Chavín and an Aymara spread with Wari, then Quechua Q.IIB/Q.IIC with Chincha during the Late Intermediate period (Cerrón Palomino 2000b). Others interpret an Aymara spread with the Early Horizon Chavín and a Middle Horizon c. 500 CE Quechua expansion in the context of the Wari state (Heggarty and Beresford-Jones 2010:Fig. 3; Beresford-Jones and Heggarty 2011:Figs. 2–3).[3]

[3] According to the most recent evidence, the linguistic domination of Quechua and Aymara has been particularly evident in the post-Conquest and Colonial periods (Heggarty 2007, 2008). The general consensus is that such language spread occurred progressively over the last two millennia and was widely replaced by Spanish (Heggarty 2008).

Although there is evidence that these language families spread throughout most of the Andes in pre-Columbian times, there is some question whether this occurred during the Early and/or Middle Horizon or Late Intermediate period. Quechua expansion is reflected by evidence of intensification of food production in intermontane highland valleys referred to as the "quichua ecozone." The Wari maintained control of food production in such valleys, and the distribution of quichua ecozones roughly correspond to highland regions of early Quechua expansion (Beresford-Jones and Heggarty 2011:19; Heggarty and Beresford-Jones 2010; see also Isbell 1974; Covey 2006).

While indigenous language families in Mesoamerica are relatively scattered and fragmented, Aymara and particularly Quechua cover very extensive blocks of continuous territory throughout the Andes (Heggarty and Beresford-Jones 2010). Such fragmentation in Mesoamerica and continuous widespread language convergence of Aymara and Quechua stems in part from geographic and environmental factors that facilitate the integration and isolation of language groups over time, somewhat analogous to what has been described by Ernst Mayr (1979:341–350) with regard to biological divergence resulting from allopatric speciation. These linguistic and cultural differences make direct comparisons of the significance of lightning to pre-Columbian cultures challenging and appear to have implications for the presence of cultural horizons in the Andes and their relative absence in Mesoamerica. The continuous linguistic spread in the Andes in part explains the widespread continuity regarding beliefs, perceptions, and rituals surrounding lightning. Recent linguistic evidence had focused on differentiating Quechua dialects Q.IIB and Q.IIC as well as estimating and evaluating the temporal and geographic spread of Quechua Q.IIB into northern Ecuador at least three centuries before Inca expansion c. 1487 CE (Torero 1974, 2002; Cerrón Palomino 2000a, 2000b, 2003; Heggarty 2007, 2008).

Pre-Columbian Andean religion involves a worship of the natural world, what has come to be called a sacred landscape (Sullivan 1987, 1988; Townsend 1992; Staller 2008a). Such cultural concepts and religious beliefs are widespread and very ancient throughout the Andes (Rowe 1946; Burger 1992; Zuidema 1964, 1982, 1989, 1995, 2008, 2011). Quechua is stated to have entered early on into the highland regions of northern Ecuador in the form of the Panzaleo culture complex in the context of long-distance exchange and came to dominate with the later arrival of the Incas (Torero 1974:80–84, 1985:371–373, 2002; Rostworowski 1975:340–342).[4] Northern highland Ecuador refers to the

---

[4] At the time of the Spanish Conquest, there were probably some thirty indigenous languages spoken in Ecuador, and of these languages, only twelve remain (Stark 1985:157). Colorado, Cayapa, and Coaiquer are still spoken in the western littoral, and a total of nine—Siona, Secoya, Tetete, Cofan, Huaorani, Shuar, Achuar, Quichua, and Zápara—continue to be spoken in the eastern lowlands or Oriente (Stark 1985:157).

provinces of Pichinche, Cotopaxi, Tungurahua, Northern Chimborazo, and Guaranda and cultures such as the Cayapa, Chimbu, Canelos, and so on (Rowe 1974). The southern highlands of Ecuador and northernmost highland Peru were non-Quechua-speaking regions occupied by the Cañari, on the west side of the cordillera and the Jívaro on the east and Palta and Malacato to the south (Rowe 1974). Lightning *huacas* in these regions would have been given different place-names. During the later Inca incursion into northern Ecuador, lightning *ushnus*, hilltops, springs, and so on were only given Catequilla toponyms in Quichua-speaking regions of northern and central highland Ecuador (Staller 2007:Fig. 2). *Huaca* (Aymara; also spelled *waká*) and *apacheta* (Quechua) refer to the sacred and extraordinary—for example, a superhuman person, place, object, or image embodying an extraordinary quality (Garcilaso de la Vega 1966 [1609]:77, 76–77).[5] In the current vernacular, *huaca* generally refers to ancient shrines or archaeological sites that have cultural and/or religious ties to indigenous Andean populations.

During Inca expansion and the early Colonial period the Catequil cult was primarily centered on lightning (Topic et al. 2002). Lightning was the major theophany of weather in Inca religion, known as Ilapa, now Illapa, the Hispanic spelling. The term *Illapa* refers to the physical phenomenon of thunder and lightning and to lightning bolts. In some regions, lightning is believed to be a manifestation of a vitalizing force or animating essence (*enqa* or *sami*), a life force believed to permeate all things (Flores Ochoa 1977; Gade 1983; Allen 1988). *Sami*, or *wasisami*, represents an "abstract vitalizing force" or "animating essence" that unites other sacred Quechua terms, such as *enqa* (Flores Ochoa 1977:229–230; Isbell 1978a:46; Allen 1988:207). The concept is reciprocal and spatial and given form and meaning by the natural surroundings, essentially by the environment, geography, and ecology (Staller 2001:26; Rostworowski 1986:31).

Spirits associated with lightning are central to Andean cosmology and religion and are depicted in the iconography of a number of cultural traditions (Gade 1983; see also Urton 1981; Salomon 1982; Salomon and Urioste 1991). Among the Inca, Illapa was regarded as the animating essence that controls thunder and, by extension, all celestial bodies and climatic forces, particularly rain, hail, and rainbows (Rostworowski and Morris 1999:792). Synonyms were Liviac (Libiac) or Chuki Illa, terms that have not survived in current Quechua usage (Rostworowski 1983:24 n.; Topic et al. 2002:308). Among the Aymara, *qhaqya* is more common than Illapa (Sullivan 1996:184). The major categories of

[5] According to Garcilaso de la Vega (1966 [1609]:73, 76–77) a *huaca* is "a sacred place" or "a sacred thing," such as "idols, rocks, great stones, or trees which the enemy [Devil] entered to make the people believe he was a god. They also gave the name huaca to things they offered to the Sun, such as figures of men [figurines and statues], birds, and animals made of silver, gold or wood."

lightning recognized among contemporary Aymara and Quechua speakers show clear Hispanic and colonial influence. In Chincheros (Cuzco Department, Peru), indigenous Andeans make gender distinctions between a *rayo* and a *relámpago*: "female lightning" or *rayo* is a flash reaching the ground, which can be lethal; "male lightning" or *relámpago* is a flash between clouds and thus not as harmful (Urton 1981:91–92).[6] While meteorological distinctions generally follow the Spanish usage, the gender dichotomy is indigenous.[7] Another type of lightning, *sullaje*, is visible as a diffuse light on the horizon and corresponds to sheet lightning. Ball lightning is so rare an occurrence that it apparently does not enter into the Andean meteorological lexicon (Gade 1983:772). Aymara speakers in Chucuito, Peru, even discriminate between *wana kaxya*, a cloud-to-cloud flash translated literally as "dry lightning," and *warawarani kaxya*, a cloud-to-earth flash referred to as "star lightning" (Tschopik 1951:226; cf. Gade 1983:772).

Among Quechua-speaking Andeans, the same cult of good luck for merchants associated with the celestial twins in Old World lore inheres in the Andean concept of *illa*, where besides meaning "thunderstone" and "twin" and forming the root of the word for lightning (*illa-pa*), *illa* also bears the definition *Illayoc runa* (literally, "a man who has *illa*"), "a rich and lucky man, one who gets rich quickly and has great luck" (Sullivan 1996:184). Significantly, *illawi* (domesticated snake; black-and-white serpent) also has reference to lightning (Sullivan 1996:185), as do felines, particularly the jaguar and, in the high Andes, the puma (Staller 2007, 2010b; Topic et al. 2002). The symbolic association with black serpents has literal reference to the electric eel (*Electrophorus electricus* L.), the only species of the genus,[8] which is exclusive to South American rain forests and the Upper Amazon. Electric eels are dark gray and swim in freshwater streams, particularly the floodplains of the Amazon and Orinoco drainages and their associated tributaries. Electric eels also range into tropical forest swamps and creeks and live on muddy bottoms in calm water or stagnant streams (Piper 2007).

The extreme verticality of the Andean environment selects for interregional economic interdependence for resources that are not available locally, or what has been termed "vertical complementarity" (Staller 2006:462–464,

---

[6] Urton (1981) states the indigenous emphasis is that *rayos* are vertical, while *relámpagos* are horizontal. Such characteristics may have some bearing on why *apu* or mountain lords are symbolized by the cross or *crucero* long before the arrival of the Spaniards.

[7] In other words, such distinctions are made by indigenous Andeans rather than by Latino populations. Gender distinctions regarding *rayos* or lightning bolts that strike the earth and *relámpago*, cloud-to-cloud flashes, may have implications for pre-Columbian religious beliefs and ritual practices with regard to Illapa, as suggested by contemporary Andean legend and folklore. For example, *rayos* do not have the power to kill or carry away males, only women (Urton 1981:91–92).

[8] Electric eels are capable of electric shocks of up to 600 volts (Piper 2007). Anyone who might have experienced such a powerful shock from an eel as well as a lightning bolt would certainly make the logical connection.

2010a:55–58; see, e.g., Murra 1975a [1972], 1975b; Morris 1979). The earliest spread of what are later described as objects and goods related to lightning rituals and religious venerations correspond with the spread of *Spondylus/Strombus* shell artifacts, maize, ceramic technology, and metallurgy in the context of early long-distance interaction c. 1500–1000 BCE (**Figures** 8a–c) (Uhle 1922, 1923; Saville 1924; Paulsen 1974; Salomon 1987; Staller 2007; Salazar 1998; Temme 2000; Rehren and Temme 1994; Bennett 1946; Collier 1946; Collier and Murra 1943; Bruhns 1994). Status-related objects made of precious metal and the crafting of metal objects appears to have ancient associations to lightning.

In the Andes, metal in general, and gold in particular, had a special status, imbued with religious significance and perceived as a symbol of power (von Hagen and Morris 1998:95). Gold was the first metal to be worked, and most metal objects throughout the Andes were made from *tumbaga*, gold-copper alloys (Uhle 1922, 1923; Saville 1924; Jijon y Caamaño 1920; Verneau and Rivet 1912; Temme 2000; Lechtman 1976, 1984, 1988; Rehren and Temme 1994). The metallurgical arts require control over firing temperatures, and kilns were often located on hilltops and cliffs, presumably to maximize exposure to diurnal changes in wind patterns (see, e.g., Uhle 1922, 1923; Jijon y Caamaño 1920; Verneau and Rivet 1912; Salazar 1998; Bruhns 1994; Bruhns et al. 1994; Temme 2000; Villalba 2011).[9] Many lightning *huacas* and *ushnus* were crafting centers and often included burials with funerary offerings of precious metal (Uhle 1922; 1923; Saville 1924; Jijon y Caamaño 1920; Lechtman 1976, 1984; Salomon 1987). Moreover, Illapa was represented as a faceless golden idol, also suggesting a relationship between lightning and the metallurgical arts. Funerary objects of precious metal were what primarily attracted the Spanish to such places, and this is why so many hilltop *huacas* were extensively looted in the early Colonial period (see, e.g., Bennett 1946; Collier and Murra 1943; Staller 2003).

The early spread of such status-related objects and funerary offerings and their associated technologies appears to have been based on traditional forms of reciprocity regarding corvée labor and resource sharing, similar to those identified during the Contact and Colonial periods (Murra 1975a[1972]; Lechtman 1984; Morris 1993; Salazar 1998; Staller 2000–2002). But because there was a history of conflict between the Cañari and the Inca, many of the local Inca installations, *huacas*, administrative centers, *tambos*, *ushnus*, and so

[9] In southern highland Ecuador, the wind patterns along the eastern side of the cordillera and in the valleys around Cuenca and Loja are highly regular and of increasing intensity during midafternoon, specifically between 1 and 3 p.m. Consequently, given the naturally occurring ore deposits, the relatively warm climate, and the irregular topography, this was a region of many metallurigcal sites. Their kilns are generally subterranean, with drafts or openings facing the prevailing winds, which of course greatly increase the kiln temperatures necessary for working metal (Lechtman 1976, 1984, 1999). Putushio is one such hilltop metalworking site (Temme 2000). With occupations extending back to the Late Formative, it has pre-Inca artificial terraces both on its summit and at the base of the hill and Inca gold, silver, and *Spondylus* figures that were left as offerings (Temme 2000; Rehren and Temme 1998).

FIGURE 8A This silver Inca figurine covered in miniature textiles fastened with a silver *tupu* (pin) and crowned with a feather headdress was discovered with a Capac Hucha burial on the summit of Cerro el Plomo, Chile. (Photo by Hugo Maertens-Bruges)

FIGURE 8B Gold female Inca figurines without accoutrements were discovered on La Plata Island, Ecuador, with a burial. (Gold/copper alloy, early fifteenth century, Collection of the Museo Antropológico del Banco Central del Ecuador, Guayaquil; photo by John E. Staller)

FIGURE 8C This gold female Inca figurine without accoutrements is an offering from the Temple of Pumapungo at Tomebamba. Many lightning *huacas* were also crafting centers involved in the metallurgical arts. (Gold/copper alloy, early fifteenth century, Museo Antropológico del Banco Central del Ecuador, Cuenca; photo by Juan Pablo Merchán; from Idrovo 2000b: cover)

on in these highland regions were destroyed by the Cañari after the assassination of Atahualpa by the conquistadores in 1533, and then later by Spanish treasure hunters (Uhle 1922, 1923; Jijon y Caamaño 1920; Saville 1924; Hocquenghem 1994; Rostworowski 1998; Staller 2006; Covey and Elson 2007).

According to chroniclers, lightning was also symbolized by a constellation outlining a man wielding a club in his left hand and a sling in his right (Sullivan 1996; von Hagen and Morris 1998:125; Rostworowski and Morris 1999:792). Their description suggests that this constellation was probably Orion (**Figure** 9a).

Among the Inca, Illapa was also perceived as dressed in shining garments that projected flashes or bolts of lightning when he whirled his sling to bring on the coming rain (Cobo 1990 [1653]:32). Staller discovered a flat, finely carved serpentine pebble eroding from the ground just below the so-called turtle stone or Piedra de Tortuga[10] while attending an Inti Raymi festival at the Inca temple complex of Ingapirca, Ecuador (**Figures** 9b–c). Ingapirca is a temple complex commissioned by Huayna Capac and located beside the Inca highway on a mountain summit northeast of Cañar. Carved on one side of the black pebble is a face with a headdress decorated with an undulating serpent (see **Figure** 9b). Southern highland Ecuador is the only area of the Andes where both gold and silver ore deposits are found within a single valley. Numerous gold, silver, and copper ore sources are found in this region of the Andes. There are numerous *huacas* and other crafting sites involved in metallurgical arts and/or the manufacture of finely crafted pottery dating back to the Formative periods (Lechtman 1976, 1988; Salomon 1987; Rehren and Temme 1994; Hosler et al. 1990).[11] The undulating serpent on the helmet is represented in the celestial realm by a dark cloud constellation called Machacuay located in the Milky Way. This steplike configuration of the serpent reflects the celestial configuration of this dark cloud constellation, which is visible in the Valley of Cuzco during the rainy season (October to April), particularly at the beginning of the season, between October and November (Magli 2005:29; Zuidema 2011). These are the times of the year when Illapa was venerated in Cuzco. One side of the artifact has triangular perforations associated with the center of the headdress and lateral perforations with ear spools. The lateral notches serve to demarcate the head from the torso, which is depicted by the hands extending together, a common pose in Andean sculpture and figurines (see **Figure** 9b). The perforation on the right side is slightly above the left perforation, very closely approximating the celestial configuration formed by the stars Alnilam, Saiph, and Rigel on the belt of Orion. Inti the sun is carved on the other side of this artifact (see **Figure** 9c). However, the triangular perforations have no reference to this carving, only to what is on the other side, the Inca portrait (see **Figure** 9b). This serendipitous discovery provides indirect evidence of symbolic associations of Machacuay to Illapa, through the undulating serpent on the headdress, and to Orion through the triangular perforations. Its discovery near the Piedra de Tortuga is intriguing for its implications of similar symbolic

[10] The turtle stone is a living rock shaped like a Galapagos turtle. Ingapirca is an imperial center commissioned by an Inca emperor and noted for its Sun Temple and Temple to Viracocha, considered sacred because of the unusually shaped rocks and cliffs as well as the surrounding waterfalls and subterranean springs. It was a Cañari *huaca* before Inca expansion into this region, as evidenced by circular buildings, which contrast with the geometric architecture of the Inca.

[11] Consequently, many of the pre-Columbian sites have been extensively mined and looted since the Conquest period, and in many cases we have very little if any ethnohistoric information on such sites, particularly those that were looted early on during the Conquest period (see Collier 1946; Bennett 1946; Collier and Murra 1943; Staller 2003).

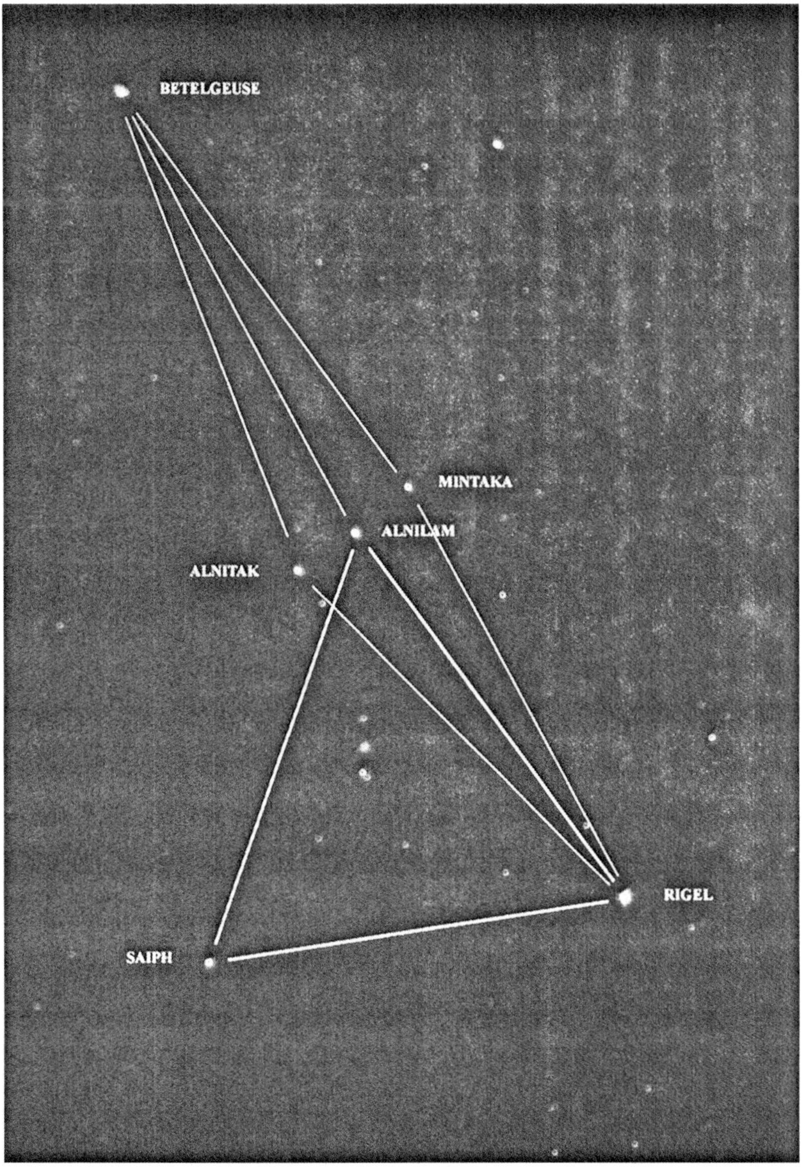

FIGURE 9A Constellation of Orion in the night sky over Cuenca, Ecuador. The belt and the celestial triangle comprising the stars Alnilam, Saiph, and Rigel are highlighted. (Photo by Natalia Biani; from Stross 2006: Figure 42–3)

associations in both culture areas.[12] This small pebble is finely carved and highly detailed, suggesting the work of an artisan probably in the metallurgical

[12] Among the Maya, the celestial triangle of Orion has reference to the three hearthstones of creation, while the three inline stars of Orion's belt are symbolic of the turtle.

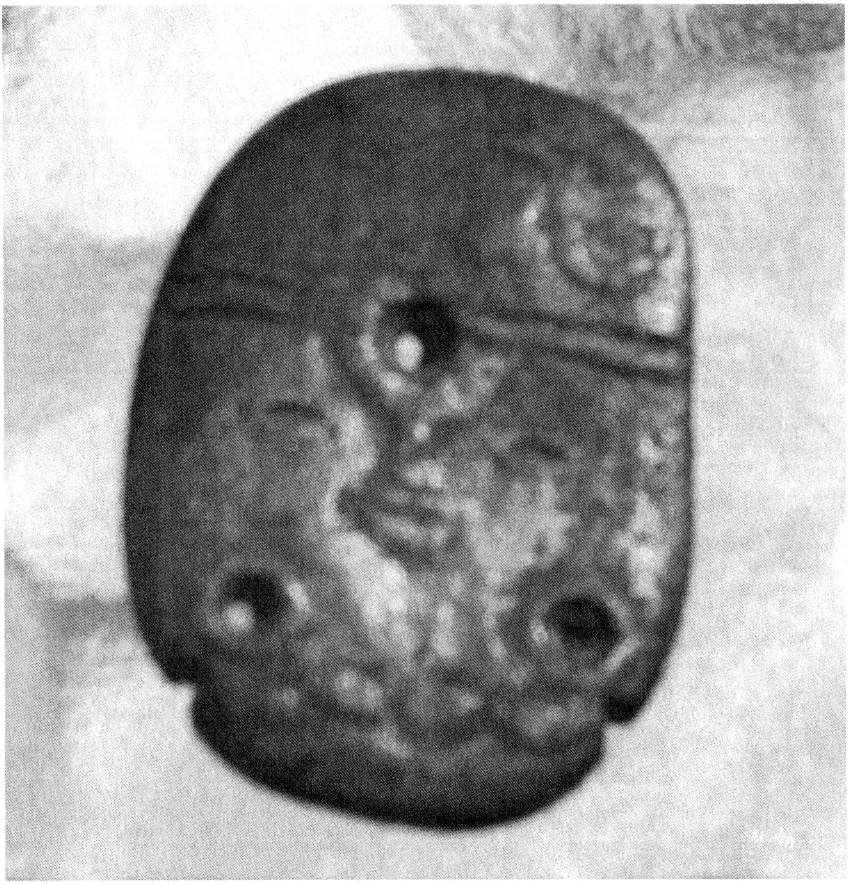

FIGURE 9B Flat, oval-shaped carved stone serpentine pebble measuring 5.5 cm. x 4 cm. x 4 cm. On one side is a face with a headdress decorated with an undulating serpent. The triangular perforations refer to the center of the headdress, the lateral perforations to ear spools. The right perforation is slightly above the left, approximating the configuration formed by the stars Alnilam, Saiph, and Rigel. The lower lateral notches provide an outline for the head. Hands extending in front of the torso is a common pose for figurines and anthropomorphic sculptures. (Photo by John E. Staller)

arts, as such carved pebbles could have been used as molds to craft or leave impressions on molten *tumbaga* or gold-copper alloy (Saville 1924; Verneau and Rivet 1912; Lechtman 1976, 1988; Rehren and Temme 1994; Hosler et al. 1990).

Factors regarding the spread of Quechua and Aymara over such a vast area of the Andes are also related to cultural responses to traditional Andean economy and adaptation to the extreme geography and ecological variability

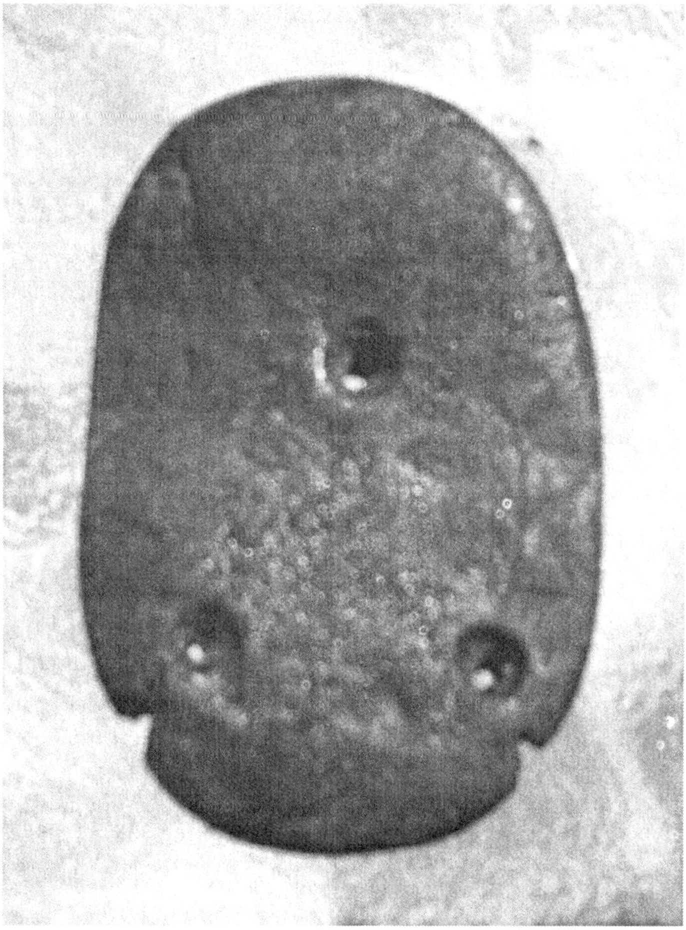

FIGURE 9C Flat, oval-shaped carved stone serpentine pebble measuring 5.5 cm x 4 cm x .4 cm, with a carving of Inti the sun on the other side. The triangular perforations and side notches have no reference to the anthropomorphic sun carved on this side. (Photo by John E. Staller)

(Murra 1975a [1972], 1985a, 1985b; Morris 1979, 1993). Archaeological evidence suggests that vast networks of roads geographically linked parts of the Andes. Scholars have documented such earlier roads in the context of Wari expansion in the Middle Horizon period. Roads also connected different regions of the coast. Certain regions of the southern highlands of Ecuador and northern highlands of Peru also appear to have road networks that predate Inca expansion (Hyslop 1990; von Hagen and Morris 1998; Covey 2006). Such road systems not only facilitated travel and communication across the Andes cordillera and along the coast, but also served as physical links to *huacas* and their conceptual links between different Andean centers, their hinterland, and support populations (von Hagen and Morris 1998:186–189). Chroniclers state that they traveled Inca roads that "cut through deep valleys and over mountains" and also "cut through rock" and had "storehouses, temples of the sun and posts [*tambos*] along the way" (Cieza de León 2005 [1553]:170–176). These road networks spanned the entire Inca Empire, from northernmost Ecuador at the imperial center of Incahuasi de Caranqui in Imbabura Province, Ecuador, extending south to present-day Santiago, Chile, a straight-line distance of almost 3,800 kilometers (2,280 miles) (**Figure** 10).

There were a number of ancient roads in the southern highlands of Ecuador, including two of particular importance: one beside the Río Cañar and another beside the Río Jubones. Such roads were involved in the *Spondylus* trade beginning in Formative times, c. 2200–1800 BCE (see Paulsen 1974; Hocquenghem 1994; Salazar 1998). The conquisator Pedro Cieza de León (2005 [1553]) mentions that Inca palace estates and their associated storehouses were found all along the Camino Real:

> I have alluded to the great power of the Kings Yncas of Peru, and to their surpassing valour, and how, along a distance of more than one thousand two hundred leagues of coast which was under their rule, they appointed their delegates and governors, and formed many deposits full of all things necessary for their troops. In some of these deposits there were lances, in others darts, and in others sandals, and so on with other arms and articles of clothing which these people use, besides stores of food.... Besides these depots and lodgings throughout the kingdom, there were palaces and temples of the sun at every ten or twenty leagues along the road, where there were priests, Mama-cunas, virgins, and more complete supplies than at the other stations. There were also governors, or chief captains, appointed by the Ynca, with the Mitimaes and Indians bound to service....The Ynca had people...to collect the tribute in their districts, and see that all necessary supplies were kept in readiness. (Cieza de León 2005 [1553]:164)

The sophistication of the engineering and labor that went into constructing these road networks and their associated *tambos* and administrative centers rivals anything in the ancient world (Pillsbury 2004; Morris 2004; Covey 2006; von Hagen and Morris 1998). The Inca highway, or Camino Real, was a visible reminder of the power and sovereignty of the Inca state and their imposition of an ethical order on the landscape. The Spaniards recognized the fact that the Inca were a highly civilized people with amazing technologies and administrative abilities. They were also favorably predisposed toward some regions under their rule, as indicated

FIGURE 10 Governors of the bridges, *governador de los puentes deste Reyno*, high-status functionaries, indicated by headdress and cloths, administered the Camino Real tracking traffic and movement of people. Guaman Poma writes, governors of the bridges were dispersed at *tambos* along the Camino Real and maintained the most extensive pre-Columbian road network in the ancient world. Bridges made of fiber cord joined steep mountain passes and ravines. (From Guaman Poma 1980 [1583–1615]: 328, fol. 356)

by this description: "In the district of Quito, not very far from Tumebamba, there is a province called Chumbo, but before reaching it there are other villages...inhabited by Indians wearing clothes, with good-looking women." Many of these villages included "royal buildings... as in those we have passed, and the people obeyed the Lords Yncas, and used the general language [Quechua] which was ordered to be talked in all parts." He later mentions that in the province of Chumbo the Camino Real "continues for fourteen leagues over rugged and sometimes difficult ground until a river is reached where there are natives with balsas who ferry travelers across. This place is called the pass of Huayna Ccapac, and it is said to be twelve leagues from the Island of Puná" (Cieza de León 2005 [1553]:171–172).[13] These ancient road networks facilitated the convergence and spread of the Quechua and Aymara language families and consequently provided a linguistic basis for similar symbolic associations and significance regarding particular *huacas* or sacred places, as well as surrounding rituals, and beliefs associated with lightning over vast regions of Andean cordillera and adjacent coastlines.

These data support previous assertions by various prehistorians that the Andes became increasingly interconnected by networks of roads well before the rise of the Inca Empire (Hocquenghem 1993a, 1993b, 1994; Salazar 1998, 2004). In Peru, such roads appear initially associated to spread with the Huari (Wari) and Tiwanaku civilization during the Middle Horizon (Kauffman Doig 1978), while in regions to the north they appear to have been related to trade in marine shell and salt very early on (Hocquenghem 1993a, 1993b). Inca expansion and the spread of Quechua and Aymara were greatly facilitated by the Inca highway, connecting mountains and valleys along the second highest mountain range in the world. The Inca highway—or Camino Real, as chroniclers called it—extended from Cuzco north to Quito and as far south as present-day Santiago, Chile, and included branches that ran along the coast as well as the cordillera (Hyslop 1990; Hocquenghem 1994; von Hagen and Morris 1998).

The ethnohistoric accounts also indicate that lightning was always linked to the landscape and ancestor veneration in Andean culture; fictive kinship relations to *Illapa ushnus* or *huacas* were common (Staller 2010a). The importance of *chicha* or maize beer to Andean ritual and political economy has been widely documented (Murra 1975a, 1975b, 1985a, 1985b; Hocquenghem 1993a, 1993b, 1994; Salazar 1998; Morris 1979, 1993; Staller 2006, 2010a). Maize beer

---

[13] It should be noted that archaeologists have identified continous *Spondylus* and *Strombus* workshops in the northern part of Puna Island dating back to the Formative periods, c. 2200–1800 BCE, and extending until the Conquest, c. 1531 (Hocquenghem 1994).

was often consumed and offered in rituals associated with lightning veneration. This is clearly brought out by the Spanish conquistadores and their encounters with local populations when they landed on the Tumbez coast. Cristóbal de Molina told Francisco Pizarro that there was a very strong fortress on the Tumbez coast; it contained six or seven stone walls, "and there were many riches inside." Pizarro considered this so remarkable that he did not believe it and sent Pedro de Candia,[14] who is described as "quite ingenious," to investigate (Cieza de León 1998 [1553]:112; see also Molina 1959 [1575]). Cieza de León (1998 [1553]) also suggests an association with the fortress and precious metal:

> When he was back on the ship, Candia told the captain so many things that what Alonso de Molina said was nothing in comparison. [Candia] said that he saw silver vessels and many silversmiths working, and that on some walls of the temple there were gold and silver sheets, and that the women they led of the Sun were very beautiful. The Spaniards were ecstatic to hear so many things, hoping with God's help to enjoy their share of it. We learned that a messenger went from Tumbez with all speed [along the Camino Real] to Quito to King Huayna Capac to report all this and to inform about the kind of people [the strangers] were and the nature of the ship. But they say that when the news arrived, [Huayna Capac] was already dead.... It is considered certain that [Huayna Capac] died the same year and time that Francisco Pizarro arrived on the coast of his land. The time that he [Francisco Pizarro] spent in the mangroves, Huayna Capac had spent doing great things in Quito. And as everything is arranged and ordered by divine permission and disposition, it pleased God that while Huayna Capac reigned and lived, although pagan, the Spaniards would not enter his land. (Ibid.:113–114)

This description suggests an Inca palace or estate. Pedro Candia made such an impression because he was one of the tallest men in Peru, and his stature and use of arms both frightened and impressed the Andeans. Despite Cieza's claim that when the Inca messenger arrived in Cuzco, Huayna Capac

[14] Pedro de Candia was considered one of the largest men in Peru and specialized in the use of firearms. Born in Crete in 1494, Candia fought against the Turks as an artillery specialist and participated in the Italian campaigns. He then married in Villalpando (Zamora) and came to the New World in 1526 on the expedition of Governor Pedro de los Rios to Tierra Firme. In Peru, he took an Inca princess as concubine, and their son was a friend of the mestizo chronicler El Inca Garcilaso de la Vega. The conquistador Diego de Almagro the Younger accused Candia of treachery and killed him at the end of the Battle of Chupas in 1542.

was already dead,[15] the Andean chronicler Guaman Poma later depicts Huayna Capac in Cuzco asking Pedro de Candia if he eats gold (**Figure** 11). Is this because the ruler believed him to be Choqui Ylla Uillca (*el noble del rayo, o de oro*), the noble bolt or gold, or because Guaman Poma was portraying Pedro de Candia as greedy? As mentioned above, the encounter documented by Cieza de León was a result of an overheard conversation between Cristobál de Molina and Francisco Pizarro regarding an Inca "fortress" along what is today the Tumbez coast. Speaking of this "fortress," which was no doubt an Inca palace estate, Cieza de León (2005 [1553]) observed that

> when it was finished Huayna Ccapac ordered a temple of the sun to be built, and two hundred virgins, from amongst the most beautiful daughters of the chiefs of the province, to be collected together in it. In this fortress which before it was ruined, is said to have been a thing worthy of notice Huayna Ccapac had his captain or delegate, with a number of Mitimaes, and stores and provisions for their maintenance, as well as for the troops that passed that way. They also say that a lion and a very fierce tiger were placed in the fortress and ordered to be well guarded. (Cieza de León 2005 [1553]:193)

In another chronicle, Cieza de León (1998 [1553]:112) recounts that when Pedro de Candia and the Spaniards landed along the Tumbez coast of northern Peru, as always "the Indians were on the beach and they went with him until they brought him unto the presence of the Lord of Tumbez," who was accompanied by his people. Noting Candía's stature, the lord and his subjects "begged him to discharge his arquebus, as he had before on his ship." When he did so, "the Indians fell to the ground, others screamed, they judged this Christian as very brave because of his stature and for discharging those shots." The Lord of Tumbez also brought a lion (puma?) and a *tigre* (jaguar), "to see if Candia could defend himself from them or they would kill him" (ibid.:112). Candia related that he again discharged his arquebus, and more Indians fell to the ground, but the felines approached him and were very gentle and did not harm him:

> They brought them and set them loose on Candia, who, having the arquebus loaded, fired it, and more Indians than before fell to the ground in fright. And without the Indians, the animals came to him as gentle as if they were lambs, as Candia told it. The cacique ordered

[15] There is much debate regarding the exact date of Huayna Capac's death. It seems the Inca died, probably of smallpox, sometime between 1525 and 1531. Cieza suggests here that it must have been after the encounter with the Lord of Tumbez and his feline, that is, after April 1528 (Rowe 1978:83–84).

FIGURE 11 *Conquista. Guiana Capac Inga, Candia Español.* The ruler Huayna Capac sits beside a plaza in Cuzco holding a plate of gold nuggets. He asks Pedro de Candia, "Comes este oro?" or "Do you eat this gold"? Candia responds, "Este oro comemos," or "This is the gold we eat." Huayna Capac was informed by a messenger of the Spaniards' arrival along the Tumbez Coast, where the Lord of Tumbez told Candia he was like the "thunder in the heavens." The ruler's question reinforces a symbolic link between lightning and precious metals. (From Guaman Poma 1980 [1583–1615]: 343, fol. 369)

> them returned to where they had been. He asked Candia for the arquebus and poured many cups of their maize wine into the barrel, saying: "Take it, drink, since one makes such great noise with you that you are similar to the thunder of the heavens." And he ordered Pedro de Candia to sit down. They gave him plenty to eat and asked him many things about they wanted to know. (Cieza de León 1998 [1553]:112–113)

In making this toast to the Spaniards, the Lord of Tumbez invokes the life force or animating essence of Illapa (MacCormack 1991:286). The deadly power and deafening noise of cannons and arquebuses presumably account for the name, *illapa*, that Andeans gave to these Spanish weapons (Cieza de León 1998 [1553]:114). The Lord of Tumbez then brought Candia to the so-called fortress described by Molina, and on their arrival, Pedro Candia's presence was requested by the Inca princesses (*mamaconas*):

> He saw the fortress. The mamaconas, who are the virgins, wanted to see him, and they sent to beg the ruler to him there. Thus it was done. They were extremely pleased to see Candia. They were skilled in working with wool, of which they made fine cloth, and they were in the service of the temple. Most of them were beautiful, and all were very affectionate. (Ibid.:113)

These "sacred virgins," as the Spaniards sometimes called them, were renowned as weavers and for their service to the ruler and the temples and *ushnus* to the sun, moon, and lightning (Zuidema 1989). *Mamaconas* played an important role in preparing *chicha* for the ruler and his entourage. In fact, only they could make *chicha* and clothes for the ruling elite (Morris 1979, 2004; Chávez 2006; Staller 2006). *Acllyas* and *mamaconas* were essentially Inca princesses. Their high status was achieved through their relationship as wives of Inca rulers. This revealing account also emphasizes the importance of maize beer (*aqha* or *chicha*), lunar cycles, and felines to pre-Hispanic lightning veneration. *Acyllas* and *mamaconas* have strong symbolic associations to the moon, reinforcing our understanding of the importance of lightning veneration to the political economy and authority of the Inca state. The reference to *acllyas* or *mamaconas* in association with lightning *ushnus* or *huacas* is also intriguing, as only imperial installations, Inca palace estates, had buildings to house these Inca princesses (Pillsbury 2004; Morris 2004; Covey 2006; Covey and Elson 2007).

During the Late Horizon period, the imperial installations were commissioned by an Inca ruler and included temples to the sun and moon and buildings to house *mamaconas* (Pillsbury 2004; Morris 2004; Morris and Thompson 1985). Inca rulers and their *panacas* had such estates or palaces, *ushnus*, *huacas*, and administrative centers built throughout their empire (see, e.g., Idrovo 2000b; Pillsbury 2004; Pillsbury and Evans 2004; Covey 2006;

Morris 2004; von Hagen and Morris 1998). Thus the fortress to which Cristobál de Molina referred was perhaps a palace estate or more probably a lightning *ushnu*. This may explain why Pizarro considered what Molina told him "remarkable" (Cieza de León 1998 [1553]:113).

These accounts indirectly suggest certain associations between lightning (Illapa) and the metallurgical arts. In Cuzco, the Inca referred to Illapa as Choqui Ylla Uillca (*el noble del rayo, o de oro*), the "noble bolt or gold." Guaman Poma's depiction of how the conquistador Pedro de Candia interacted with the emperor Huayna Capac reinforces this symbolic association of Illapa with the metallurgical arts (see **Figure** 11). Such associations are consistent with its iconic representation in Cuzco: the golden idol without a face (Choqui Ylla Uillca), which stood in the Golden Enclosure, the Coricancha, or, as the conquistadores called it, Templo del Sol.

Lightning was associated with certain *huacas* and *ushnus* in various regions of the Andean highlands and along the coast. Early accounts mention that Catequil was a Quechua term for the thunder and lightning deity related to the *huaca* of Huamachuco (Betanzos 1996 [1551]), while later chroniclers claim it referred to the *huaca* of Cajarmarca as well as Huamachuco (Albornoz 1989 [1584]; Sarmiento 1907 [1572]), or even a *huaca* worshipped from Quito to Cuzco (San Pedro 1992 [1560]:174) (see **Figure** 6). The term *Catequil*, which was transformed to *Catiquillay* or *Catequilla* (the more common form in northern Ecuador), is now believed to have originally been a Culle term (Topic et al. 2002:306). Chroniclers emphasize two superstitions: one involved the term *Catequilla* and lunar eclipses,[16] while the second involved beliefs surrounding the association between lightning and twin births (Topic et al. 2002:306). Lunar associations are also inferred from the repeated presence of thorny oyster (*Spondylus* spp.) in ritual practices involving Illapa. Such practices appear to be ancient in this region and in the Andes in general. Albornoz and Sarmiento also refer to Catequil as the *huaca* of both Cajamarca and Huamachuco, rather than just Huamachuco, although Albornoz (1989 [1584]) emphasizes that the *huaca* of the *indios guamachucos* was "one of the most important in the Inca Kingdom." The combining of Huamachuco and Cajamarca into a single unit reflects the Inca administrative organization Hanan/Hurin, "upper/lower" (Cieza de León 1984 [1553]:226, 234; Pizarro 1978 [1571]:220–210; Zuidema 1980, 1982, 1989, 2011; Topic 1992, 1998), which was adopted by the colonial government when the *corregimientos* were set up in

[16] Topic et al. (2002:307–308) note that chroniclers mentioned Catequilla can also be translated as "follower of the moon," this with reference to the original Culle term. However, the Culle language is essentially unknown linguistically. Nevertheless, some lightning *huacas* or *ushnus* were used for astronomical purposes that may have involved recording lunar cycles or eclipses. Moreover, some Inca estates also included a house to the *mamaconas* supporting such a symbolic association (see, e.g., Cieza de León 1998 [1553]).

1565–1566 (Topic et al. 2002:306–307). During the turn of the seventeenth century, chroniclers briefly mention a *huaca* called Cati Quillay, which is described as an emissary of the Inca (Salomon and Urioste 1991:100–101) or as having been given to the community of Llacsa Tampa in Huarochiri by the Inca emissary (Taylor 1987:293). Although this document is clearly based on local rather than Cuzco Inca informants, it is possible that the term *Cati Quillay* may represent an Aymara variant of *Catequil*, as the Huarochiri region was originally Aymara and adjoins Yauyos Province, where Central Aymara is still spoken to this day.

Verb tenses imply that an idol dedicated to Cati Quillay was temporarily located in the community of Llacsa Tampo and credited with forcing one of the fictive sons of the oracle at Pachacamac, Llocllay Huancupa, to speak (Topic et al. 2002:307–308). The importance of this brief description is that Catequil was not just a *huaca*, or even an oracle, but a *yañca*, an oracle that could make other *huacas* "speak" (Rostworowski and Morris 1999:796; Topic et al. 2002:306–307). The only other *huaca* said to have this power was the oracle of Pachacamac (Rostworowski and Morris 1999:793–794; Staller 2008b:295).

Pachacamac was one of the most powerful and influential of all oracles, second only to the Coricancha or Golden Enclosure in Cuzco in magnitude, devotion, authority, and richness during the Late Horizon period (Cobo 1990 [1653]:89; Rostworowski and Morris 1999:793; see also Kauffman Doig 1978). The oracle at Pachacamac was believed to foresee earthquakes and their intensity, and it may be that this is why it was a *yañca*. Earthquakes generally follow fault lines. The indigenous perception would have seen major quakes moving from one valley to another as empirical indications that the various *huacas* were "speaking" to one another through their vitalizing force or animating essences.

The appearance of distinct ceremonial centers and modifications of "sacred places" in the natural environment, commonly referred to as *huacas*, were the focus of veneration to Illapa (Espinosa Soriano 1988a; Staller 2007, 2008b, 2010b; Topic et al. 2002). *Huaca* refers to sacred and extraordinary things, but can have a number of meanings: it is generally used by scholars to refer to shrines or sacred places or built landscapes modified by architecture, rock art, artificial terraces, and so on (see Garcilaso de la Vega 1966 [1609]:76–77). Lightning (Illapa) has mythical dimensions in the cosmology and clear symbolic associations to epochs of destruction and creation in Inca cosmogony. Illapa tends to have a figurative meaning. *Qhaqya*, the Aymara term is also used by Quechua speakers in southern Peru, can also refer to the malevolent spirit in lightning. *Lliwlliy* (Quechua) and *qhejsiril* (Aymara) are two other terms for lightning (Gade 1983:770). These associations and the surrounding terminology in both language families has a relatively wide geographic distribution compared to Mesoamerica, and our research suggests that this may be related to the spread of Quechua and Aymara before Inca expansion.

Lightning in the Context of Pre-Hispanic Andean Religion

Nature's wrath is a recurring and unpredictable reality in the Andes mountains, particularly the frequent and spectacular electrical storm (*salla*) in the altiplano of Peru and highland Bolivia. Most thunderstorms in the Northern and Central Andes occur between December and February. They are common along the western cordillera and particularly in the puna and altiplano of the high sierra. It brings diluvial rain (*para*), furious wind (*wayra*), destructive hail (*chikchi*), thunder (*sallallaya* or *qhaqya*), and, most violent of all, lightning (*illapa*) (Gade 1983:770). The sanctification of the natural world and cultural concepts and beliefs surrounding the landscape are ancient in Andean culture. Scholars have emphasized the importance of religious cults to the early spread of ancient technologies, ethnogenesis, or cultural horizons (Willey 1955, 1962, 1999; Burger 1988, 1992; Heggarty 2008; Heggarty and Beresford-Jones 2010; Beresford-Jones and Heggarty 2011).

Pachacamac was a major pilgrimage center, a monumental complex of temples, plazas, storehouses, and elaborate burials—a necropolis with many elite tombs from the entire pre-Hispanic sequence overlooking the Pacific Ocean (**Figure** 12a). It was one of the most feared as well as most revered pan-Andean oracles, and the height of its religious influence and power dates to 800–1000 CE (Kauffman Doig 1978; von Hagen and Morris 1998). Early occupations date to about 100 CE, and the archaeological evidence suggests that the oracle was highly venerated throughout the Andes long before the rise of the Inca Empire (von Hagen and Morris 1998:155–157). Oracles, *huacas*, and divination played a major role in Inca culture, particularly in interacting with the natural environment and surrounding landscape (Guaman Poma 1980 [1583–1615]; Hyslop 1990; Staller 2006; Topic et al. 2002). There is evidence to suggest that such centers and sacred places played a similar role going back to the very beginnings of Andean civilization (Hyslop 1990; von Hagen and Morris 1998).

The importance of weather and celestial essences or deities is apparent in the Titicaca Basin during the Middle Horizon period in the Tiwanaku civilization (**Figure** 12b). One of the most striking carved stone monuments associated with this civilization is the Gateway of the Sun. The central figure over the portal is referred to as the Gateway God or Staff God, called Thunupa and worshipped throughout the Titicaca Basin for its power to bring rain, thunder, and lightning. Scholars have suggested that Inca religious cosmology to some extent modeled Illapa after Thunupa (von Hagen and Morris 1998:124–125). However, where the sculptures at Tiwanaku combined solar-related imagery with their weather-related

FIGURE 12A Pachacamac, Lurin Valley, central coast of Peru. Aerial view of the oracle center in the early part of the last century. Pachacamac was partially restored, but in recent decades the outskirts of Lima have slowly encroached into the surrounding landscape. Pachacamac was venerated throughout the Andes, of such religious importance. An Inca emperor commissioned a Temple to the Sun (Inti), the large structure on the promontory, at this religious center. (Photo by Shippee-Johnson, courtesy of Library Services, American Museum of Natural History; from von Hagen and Morris 1998: Illustration 103; negative 334841)

deity, the Inca made lightning, along with the sun and moon, part of their primary focus of religious veneration.

The religious importance of lightning and the spiritual essences that it embodied is evidenced by the fact that the idol was kept in the same sanctuary as the idols to Quilla the moon and Inti the sun (Hemming 1970:132–136; Gade 1983:775; Staller 2007, 2008b). When a place in the landscape was transfigured by a lightning bolt, it became a *huaca* or *Illapa ushnu* (Gade 1983:776). The term *ushnu* infers such *huacas* were perceived as seats of power associated with administering the empire. *Ushnus* are also characterized and distinguished by architectural features such as vertical openings or holes in which libations and ritual offerings were made (Zuidema 1980, 1992, 2011). Many had astronomical functions and used cut-stone canals to channel fluids or were essentially *axis mundi* in that they provided basis for conceptualizing the surrounding landscape, the heavens, and the world below (Zuidema 1983, 1989,

FIGURE 12B Gateway of the Sun at Tiwanaku. The Gateway or Staff God over the center of the portal may represent the weather- and sky-related deity Thunupa. The Gateway God was worshipped throughout the Titicaca Basin to bring rain, thunder, and lightning. (Photo by Wendell C. Bennett, courtesy of Library Services, American Museum of Natural History; from von Hagan and Morris 1998: Illustration 82; negative 600909)

1995, 2011; Staller 2008b; Meddens et al. 2008). Some *huacas* were located at rock outcrops or centered on large boulders that had been cleft by lightning (Arriaga 1968 [1616]:205–231). These places were often walled in by the Inca to emphasize their sanctity (Arriaga 1968 [1616]:205; Gade 1983:776).[17] Pre-Hispanic lightning *huacas* and *ushnus* are usually on high hills, on cliffs, or near mountain passes and water sources (underground springs and/or waterfalls). Ritual offerings commonly found at such *huacas* and *ushnus* include worked and unworked *Spondylus*/*Strombus* shell, sling stones, or river-rolled pebbles, sometimes covered in ocher, and evidence of maize beer (*aqha* or *chicha*) or drinking vessels (*queros*) (Staller 2007: Table 1; Topic et al. 2002:326).

[17] A more specialized lightning *huaca* was Illawasi, described by indigenous societies in the puna of the Huaylas region, at which sacrifices were made to ensure herd maintenance (Hernández Príncipe 1923 [1622]:27).

At the time of the Conquest in the early sixteenth century, the Inca ruled over the largest empire in the New World. Their expansion beyond Cuzco occurred around the mid-fourteenth century (von Hagen and Morris 1998:164). Within a century the Inca Empire extended over four thousand square kilometers, incorporating over much of modern Peru, Ecuador, Bolivia, northern Chile, and northwestern Argentina (see **Figure** 6). Between 1438 and the Conquest in 1532 the Inca constructed over one hundred ceremonial platforms and shrines (*villcas*), most on the summits of the highest mountains in the empire (Ceruti 2004:104; Reinhard 1983, 1985; Besom 2010). Recent research has indicated that some Colonial period *huacas* or shrines (*villca*) were associated with the lightning cult Catequil. The Aymara term *villca* can mean both "sun" and "shrine" (Bertonio 1956 [1612]:386). In the Huarochirí manuscript, it means someone who has entered into the society of *huacas* by achievement or marriage—a very important cacique, or chief (Zuidema 1973:19; cf. Salomon and Urioste 1991:46). Salomon and Urioste (1991:46) suggest that *villca* refers to a person who partakes of the status of a *huaca*, a superhuman person (see also Condori Mamani and Quispe Huamán 1996).

Catequil religious veneration was symbolically and metaphorically associated with the natural world, particularly mountain summits, mountain passes, subterranean springs, and caves.[18] Catequil was sometimes given the title *apu* or "mountain lord" (Albornoz 1989 [c. 1584]:210; San Pedro 1992 [1560]:176; Topic et al. 2002:310–311). *Apu* (Quechua) and *auki* (Quechua) refer to mountain lords or spirits and are sometimes used synonymously. *Achachila* (Aymara), *mallku* (Aymara), and *wamani* (Quechua) refer to ancestors or grandfathers who are generally associated with mountain summits or mountain spirits (Sharon 2001). Mountain lords or *apu* were symbolized by a cross or *crucero*, which was often placed on the summits of sacred mountains or at high mountain passes (*apachitas*). Earth shrines or *wamani* passes are places situated on ridges and mountaintops near a trail. Andeans built crosses or placed stones or rocks at such locations in recognition of the animating essences believed to reside there; the crosses were thought to energize or restore the energy one has lost when climbing to such places (Valderrama and Escalante 1977). Bastian (1978:69) states that these *apachita* are the highest places on the ascending road, and when reaching there the Indians would discard their coca quids and throw several stones on a pile, creating a sacred cairn. Such telluric practices in Andean religious belief are a conscious symbol of the removal of one's load and the restoration of strength. Thus travelers and

[18] Archaeologists have studied Inca *huacas* with regard to architectural and geographic associations and found that about 29 percent were near springs or water sources, 29 percent near standing stone, and 10 percent near *apachitas* or mountain passes; 9 percent had temples or architecture of cut stone; and 9 percent were in fields or flat places (Magli 2005:24).

traders often placed crosses on rock piles or cairns (Condori Mamani and Quispe Huamán 1996:152). Lira (1982) provides a more general definition of *apachita* as rustic tomb or holy earth, a mound of rocks over provisional burials made for those who die while traveling. The word *apachita* is derived from the verb *apachiy*, "to send" or "to ship"—or, as Bastien (1978:69) puts it, "to have something or someone carry the load away." Mountain passes and such earth shrines conceptually link powerful spiritual forces or essences such as lightning to those who transverse the extreme Andean landscape (Condori Mamani and Quispe Huamán 1996:152 [35]).

Crosses or *cruceros* also have reference to the geography and surrounding landscape. The *crucero* symbolizes where the principal divisions of celestial (*janan pacha*) and terrestrial (*cay pacha*) space are sometimes defined in accordance with topographic features. Urton (1981:40–43) provided ethnographic evidence to indicate that the *crucero* or center marks the movement of water across the landscape, and the village of Misminay was linked to surrounding communities higher and lower in altitude. Urton also found that the crossroads near Misminay was characterized as a kind of oracle or omphalos, but on entering the circular ancient ruins of a nearby ancient ruin, one is conceptualized as moving into an underworld, or to an internal or interior place. Therefore, all of the primary cosmic layers, the celestial, terrestrial, and underworld are accessible from the *crucero*, suggesting that such places can also represent an *axis mundi* or "center of the world" (Eliade and Sullivan 1987). Even the name, *crucero* or cross, is of some significance to native Andean religion through an association with animating essences or vitalizing force (*wamani* or *apu*) on mountain summits and certain mountain passes. Such natural associations commonly linked such places to Illapa and the rituals and rites surrounding lightning veneration; they also explain in part why the cross was so readily integrated into colonial and postcolonial religious syncretism among indigenous Andean populations and why the cross of the Order of Santiago became such an important honor and symbol of Spanish conquest.

In modern Quechua, Catequilla has come to refer to overlapping terms associated with mountain essences that have masculine attributes (*wamani*) and to fertility—sources of water and streams (Isbell 1978a:164; Allen 1988:41–49; Gose 1994:299). Thus the description given to "apu Catequil" has overlapping connotations that refer to mountain spirits, thunder and lightning, and "fathers" as well as to founding ancestors and to cliffs/mountains (Allen 1988:41; Gose 1994:209, 223; Topic et al. 2002:310–311) (**Figure** 13; see color plate section). Mountaintops have a natural association with thunder and clouds in Andean culture: their significance to lightning veneration as *apu* or mountain lord has direct reference to the male generating principle, since mountain summits are the sources of streams, and thus to the channeling of fluids, the source of the earth's fertility and fecundity (Classen 1993; Staller 2008b).

Illapa is not necessarily localized in any specific place, but it may be associated with mountain summits, passes, and high hills that attract lightning. Catequil also had mythological significance and was credited as a culture hero or founding ancestor who dug the Indians of Huamachuco out from their *paqarina* (San Pedro 1992 [1560]). Thunder and lightning are still believed to cause twin births and thus have a strong symbolic and mythological association to twins, the principles of rebirth and renewal cycles, and, particularly, eclipses of the moon (Staller 2007; Topic et al. 2002). Its close symbolic association with mountain summits provided another basis for the syncretism of Christian religions with those of indigenous cultures. Such conversion and syncretism was most evident in the rituals practices and beliefs surrounding twin births and children born with particular disfigurements.

Lightning Bolts, "Sons of Lightning," and Immaculate Conceptions

The persona of Santiago as an aggressive defender manifested itself in the Andes during the 1535 Cuzco uprising by Manco Inca. Chroniclers state that the Spaniards invoked Santiago during a difficult point in the battle. Santiago is said to have descended from the heavens on his white *córcel*, killing some Indians and frightening others into submission (Acosta 1962 [1589]:373; cf. Gade 1983:778). His miraculous revelation established Santiago as patron saint of New Spain. The rulers of Spain turned Santiago into the major symbol of European military domination and cultural imperialism in the Andes. The Crown made knighthood to the Order of Santiago one of the highest honors that could be bestowed on an individual. This is evident when considering that only four of the original conquistadores of Peru, including Francisco and Hernando Pizarro, achieved membership in the Order of Santiago (Gade 1983:777). This association may also have influenced the Spanish cult to Santiago (St. James), patron of Spain, who was symbolically associated with thunder and lightning, more specifically the thunderbolt, emphasizing the sonic expression. The tenuous link of Santiago to thunder came to extend to lightning, for where a thunderclap is heard, a lightning flash has inevitably preceded it (Gade 1983:777; Platt 1997:199).[19] The Red Cross of St. James or Santiago became emblematic of the Spanish conquest of the New World and was worn by members or knights of the Order of Santiago (**Figure** 14). Pope Alexander

[19] Once transformed from a humble disciple of Jesus into Santiago, an avenger ready to do battle, St. James fit the spirit of the New World Conquest. The venerable rallying cry was "Santiago! Santiago!" The iconography of Santiago as a sword-wielding warrior mounted on his fiery steed—his most terrifying weapons were lightning (*glurya scintilla*, the Flash of Glory), the thunderbolt, and arquebus (Gade 1983:778; Platt 1997:119–200)—are Spanish in origin and not biblical. Inspiration for this tableau appears to have come from ancient Celto-Iberian and Roman tradition (Gade 1983:778).

III founded the Order of Santiago in 1171, and the Cross of Santiago was particularly popularized among the Spanish aristocracy during the reign of Philip IV between 1621 and 1665 (Madariaga 1947; Cardinale 1983).

Lightning is still is a primary agent of transformation and known to transmit the decisions of major spirits (Valderrama and Escalante 1977). Single-sex twins, considered "divided" by a thunderbolt, had a symbolic significance that a male-female set did not possess (Mariscotti de Garlitz 1978:366). An individual conceived or born during a thunderstorm was ipso facto a child of lightning and also a candidate for ritual sacrifice in the Capac Hucha rite (Cobo 1990 [1653]:224; Krappe 1932; Duvoils 1976). Victims of the Capac Hucha rites were usually buried in a special place, the *chucicancha* (Molina 1959 [1575]:94). When individuals with such abnormalities or male twins died, they would be put into ceramic urns and kept at *Illapa ushnu* or *huacas* (Gade 1983:776; see also Arriaga 1968 [1621]:205; Garcilaso de la Vega 1966 [1609] Part I:76–77). These human remains (*Illapa huacacuna*) formed the primary focus of cultic veneration wherever they were placed, most commonly on high mountain summits (Reinhard and Ceruti 2010). Protective rites made at a rural *huacas* ranged from regular libations of *chicha* to periodic sacrifices of llamas, guinea pigs, and children (**Figure** 15). Living male twins or boys with harelips or cleft palates were also referred to as "sons of lightning," or *Illapa huacacuna*, and were central to cult veneration and given protective rites and regular offering of *chicha* and sacrifices or *pagos* of guinea pigs, llamas, alpaca, and textiles during the Late Horizon and early Colonial periods (Gade 1983:776; Krappe 1932; Duvoils 1976; Reinhard and Ceruti 2010).

Another reason that male children were exclusively bestowed such ascribed status is revealed in the surrounding legend and customs: According to traditional folklore and legend, female *rayos* or lightning bolts (*intiallapa*)—those that strike the earth—affect pregnant women outside in a lightning storm by splitting the womb, resulting in twins or giving the infant deformities such as harelips or cleft palates, which are presumably empirical indicators of woman affected by the animating essence of lightning. According to traditional culture, lightning bolts or *rayos* enter the womb, causing such deformations as well as twin births, and when they result in male offspring, it has come to be perceived as a kind of immaculate conception. The significance of gender associations to *rayo* in indigenous culture represent a transcendence of a gender duality, male/female.[20] Seen another way, the *female bolt* or *rayo* + *pregnant female* = *male same-sex twins*. Male twins transcend the gender boundary. If the births result in female offspring, they do not transcend the gender boundary,

[20] The extent to which similar concepts existed in pre-Columbian Inca culture is unknown, but the gender association of different kind of lightning is intriguing for its implications for more ancient practices and beliefs.

FIGURE 14 *Las Meninas* (*Maids of Honor*), showing a close-up of Diego Velázquez, perhaps the most renowned of all the Spanish imperial court painters, with the Cross of Santiago. Philip IV, king of Spain, awarded Diego Velázquez the Cross of Santiago in 1659. Legend has it the Crown painted the Cross of Santiago on his chest. (Oil on canvas, c. 1656, Museo del Prado, Madrid)

FIGURE 15 The idols and *huacas* of Collasuyu, "Idolos Huacas, De Los Collasuios." An Aymara couple makes *pagos*, or ritual offerings, to the mummy of a venerated ancestor buried in the *chucicancha*, a special place on the mountain summit. Guaman Poma wrote "carnero negro" on the llama being sacrificed, indicating that it is black. Black and white are often associated with rituals and rites surrounding Illapa. For example, only black-and-white maize varieties were used in Capac Raymi rituals and rites that focused on penitence and blood sacrifice. The headdresses suggest lunar veneration, common to non-Quechua-speaking cultures along the western cordillera and coast. (From Guaman Poma 1980 [1583–1615]: 244, fol. 270)

thus were not afforded special status. When a pregnant woman was believed to be affected by a bolt of lightning or *rayo*, the "impregnation" by the animating essence of Illapa superseded that of the biological father. Thus only same-sex male twins were bestowed special status as "sons of lightning," or *Illapa huacacuna*. After the Conquest, such "sons of lightning" were perceived as being conceived immaculately, or without original sin; these male infants had been born to females touched by divinity—in this case, a lightning bolt. The gender-related perceptions of different kinds of lightning provide a rationale for why during the Late Horizon and Colonial periods only male twins or individuals with such disfigurements were given special status by the Inca state, and there is archaeological evidence to suggest such beliefs have earlier origins. Such cultural distinctions may also provide a basis for explaining why male twins were considered *illayoc runa*, or rich and lucky.

Such later perceptions have been syncretized so that they are more clearly consistent with Catholic ideas of the Immaculate Conception, that is, conception through a divinity thus free of original sin. Such Christian beliefs and perceptions would have been emphasized by Catholic clerics in their quest to convert the local indigenous populations. Cobo, Molina, Sarmiento, and many others roundly condemned Capac Hucha child sacrifices and emphasized the injustice that female infants and young adults endured in the context of such rituals and rites. In 1655, Philip IV commanded the knights of the Order of Santiago to defend religious belief in the Immaculate Conception of Mary (Cardinale 1983). The Augustinians in Spain and the New World also took up this cause, as it was this monastic order that is said to have first brought the holy relics of St. James to their monastery at Uclés in the twelfth century (Madariaga 1947; Topic 1992, 1993).[21] Focus on the Immaculate Conception of Mary by the Spanish Crown may have been intended, at least in part, to convert Andean indigenous populations and transfer the cult from Illapa to Santiago. Augustinian monks in southern highlands around Cuenca, Ecuador, were central in converting beliefs surrounding Illapa to Catholic orthodoxy, including the cult of Santiago (San Pedro 1992 [1560]).

> They all believe the sun to be god, but that there was also a Creator of all things, whom, in the language of Cuzco, they call Huira cocha [Viracocha]. Although they now have this belief, they formerly worshipped trees, stones, and the moon, being prompted by the devil, our enemy, with whom some of them converse.... Already in

[21] The remains of St. James currently reside in Santiago de Compostela in Galicia, Spain (Madariaga 1947; Cardinale 1983). Pilgrims from all over Europe come to the cathedral at Santiago de Compostela in northwestern coastal Spain to worship at the shrine of St. James on his feast day, July 25 (Cardinale 1983).

> many places where he was esteemed and venerated, he is now detested, and the temples of the accursed idols are destroyed, insomuch that there is no sign of an image, and many Indians have become Christians. There are now few villages in Peru without a friar or clergyman who teaches the people. (Cieza de León 2005 [1553]:162–163)

Cieza de León made these observations about these regions of the Andes cordillera less than two decades after the Conquest. There is archaeological and ethnohistoric evidence to suggest that lightning veneration has ancient origins in these non-Quechua-speaking regions of the Andes (Staller 2007; see, e.g., Cieza de León 2005 [1553]:170–176). These regions are renowned for their indigenous pottery and crafting in the metallurgical arts. Many sites in southern highland Ecuador and northern highland Peru included cemeteries with the tombs of elites and of artisans, which often featured funerary offerings made of precious metals. Consequently, most pre-Hispanic sites found to contain such riches were looted and mined during the Colonial and post-Colonial periods (see Salomon 1987).

Since the southern highlands of Azuay and Loja Province constitute those regions of the Andes under the influence and jurisdiction of the Augustian order in the early fifteenth and sixteenth centuries, most of the ethnohistoric information regarding architectural modifications and characteristic associations, particularly funerary and ritual offerings associated with Illapa *huacas*, comes from the early chronicles and later accounts of archaeologists (San Pedro 1992 [1560]; see also Uhle 1922, 1923; Saville 1924; Jijon y Caamaño 1920; Salomon 1987; Staller 2007; Topic et al. 2002). Most of the Catequil- or Illapa-related *huacas* were located in these regions of the cordillera, but most of the ethnohistoric information regarding rituals and rites associated with lightning or Illapa came from either the Inca capital (Cuzco) or clerics and colonial officials in the northernmost highlands and southern highlands of Ecuador (Cieza de León 2005 [1553]; Molina 1959 [1575]; Cobo 1990 [1653]; San Pedro 1992 [1560]). Subsequent archaeological research in these regions suggests that many Illapa *ushnus* and *huacas* identified in these regions pre-date Inca expansion and therefore may reflect earlier lightning veneration (see, e.g., Uhle 1922, 1923; Bennett 1946; Collier 1946; Murra 1946; Collier and Murra 1943; Salazar 1998; Bruhns 1994; Bruhns et al. 1994; Temme 2000; Rehren and Temme 1994; Staller 2003).

The city of Cuenca (Santa Ana de los Cuatro Ríos de Cuenca) is the capital of Azuay Province, Ecuador (**Figure** 16). The Cuenca Valley is bowl-shaped and distinguished by a confluence of four streams—the Río Tomebamba, Yanuncay, Tarqui, and Machangara—at about 2,500 meters above sea level (Cieza de León 2005 [1553]; Molina 1959 [1575]; González Suárez 1878, 1892). During the Conquest, the region was known for two *huacas*, "one called

Cañaribamba and the other Hatuncañari,[22] whence the natives and their province took the name of Cañaris, as they are now called" (Cieza de León 2005 [1553]:162). The Cañari were distinguished from other indigenous populations at this time by their headdresses (Gonzalez Suárez 1878). Cieza de León (2005 [1553]:162–163) also mentions that the Cañari elite all spoke "the language of Cuzco [Quechua]" and all "believe the sun to be god" and that they "formerly worshipped trees, stones, and the moon." Andean societies in the Cuenca Valley and regions to the north and south are renowned for their ceramic technology and textiles.[23] Early chroniclers mention that many camelids ranged to this part of the Andes: "In former times, before the Spaniards gained this kingdom, there were great quantities of sheep in these mountains . . . and a still greater number of huanaco and vicuñas. But the Spaniards have slaughtered so many, that now there are scarcely any left" (Cieza de León 2005 [1553]:163). Sebastián de Benalcázar and his soldiers conquered and waged war in some regions in this part of the Andes before chroniclers such as Cieza de León and the Spanish conquistadores arrived in 1544 (Ramos Gómez 1988; Gonzalez Suárez 1878). However, what made the Cuenca Valley and surrounding regions famous as well as infamous in the Colonial period was the presence of ore deposits, both gold and silver, as well as precious jewels, particularly emeralds, throughout these regions of the Andes. Southern highland Ecuador contains numerous *huacas* involved in the metallurgical crafts (Cieza de León 2005 [1553]; San Pedro 1992 [1560]). Cieza de León (2005 [1553]) describes the southern highlands of Ecuador in terms of its warmer climate and natural riches:

> The province of Cañaris is very broad, and full of many rivers, in which there are great riches. In 1544, they discovered such great and rich mines in these rivers, [and] extracted more than eight hundred thousand pesos of gold. The quantity of this metal was such, that they often took out of the troughs more gold than earth. I affirm this, because I spoke with a man who had taken more than seven hundred pesos of gold out of a single trough; and besides what the Spaniards got, the Indians took an unknown quantity. (Cieza de León 2005 [1553]:163)

---

[22] Hatun-cañari is considered the largest and best-built Cañari temple center in what Cieza de León called "the province of Quito." The Cañari architecture at Hatun-cañari terminates on the slope of a mountain as a high, thick stone wall. An oval tower containing two chambers form the center of the constructions; their walls are full of niches, and the outer walls are very thick, with ramparts around the inner sides (Cieza de León 1998 [1553];162 n.).

[23] Andean potters and metallurgists in the Cuenca Valley continued to sell cooking pots or ollas in local indigenous markets until the late 1990s. Metal cooking pots have now replaced most of the earlier ceramic cooking pots, but Andeans continue to sell their jewelry and textiles at indigenous markets throughout the country and other areas of Latin America.

FIGURE 16 La Ciudad de Qüenca. Cuenca was the colonial capital of southern highland Ecuador. Guaman Poma de Ayala depicts a fountain in the middle of the plaza and two women holding water jars (*aríbalo*) on their heads. Indigenous communities in Cuenca and the surrounding valleys indigenous cultures were renowned for their ceramic technology as well as crafting in the metallurgical arts. Below the illustration he wrote, "Aqui tubu grandes casas de Guayna Capac Ynga," or "Here were the large houses of Huayna Capac Inca," in reference to Inca administrative centers like Tomebamba and Ingapirca. (From Guaman Poma 1980 [1583–1615]: 926, fol. 997)

Many early Spanish occupations in this province were mining camps set up by treasure hunters. A Spanish mining camp remained in the valley of Gualaceo from 1534 until 1549, but the valley of Gualaceo was absent from the colonial records until July 1757, when it was elevated to the ecclesiastical parish of Cuenca (Salomon 1987; Salazar 1998, 2004). Consequently, much of what was taken from the ore deposits or looted from ancient *huacas* went unreported to colonial officials. The Cañari and other societies in this region were to varying degrees sympathetic to the conquistadores because they drove out the Inca. Indigenous societies in this part of the cordillera resisted using Quechua as their lingua franca, and numerous conflicts occurred before the arrival of the Spanish, in the context of Inca expansion by Tupac Yupanqui and later Huayna Capac (Cieza de León 1998 [1553]). Perhaps the most widely documented conflict with the Inca had to do with the civil war between Atahualpa and Huascar (Betanzos 1987 [1551]; Sarmiento 1907 [1572]; Molina 1959 [1575]). The Cañari sided with Huascar, and many were slaughtered by Atahualpa's troops when they entered Cuzco. The Cañaris were later greatly resistant to Spanish colonial governments as well as conversion to Catholicism (Sarmiento 1907 [1572]; Betanzos 1987 [1551]; Molina 1959 [1575]; Albornoz 1989 [1584]; Garcilaso de la Vega 1963 [1609]).

In part because of the long history of conflict between the Cañari and the Inca, many of the Inca *huacas* and *ushnus* or temple centers, and particularly royal estates, were alleged to have been destroyed by local indigenous populations immediately after the Spanish conquest. However, many such *huacas* were also looted and destroyed by Spaniards who sought treasure and riches (Cieza de León 2005 [1553]; Molina 1959 [1575]; San Pedro 1992 [1560], 1560; Albornoz 1989 [1584]) and in many cases did not wish to share the spoils with the Crown or Holy See. During his regional survey of the Río Jobones and excavations at numerous sites, including the great Inca administrative center of Tomebamba, Max Uhle (1923) observed that a great number of pre-Columbian sites in the region were *cerros*, or hilltops with artificial terraces, that were also metallurgical crafting centers. Such sites often included cemeteries that included elite burials with sumptuary offerings (Uhle 1922, 1923; Idrovo 2000b; Salazar 1998, 2004; Salomon 1987). The mountains of Tari, the hill Cohjitambo near San Juan, the hills around the ancient necropolis at Sigsig, Chordeleg, and other sites,[24] many of which were originally Cañari *huacas*, were involved in metallurgy or ceramic manufacture. Some of these *huacas* were so highly venerated that the Inca constructed buildings on them and modified them in

[24] Uhle (1922) mentioned other hilltop sites, many of which included artificial terraces and what may have been Capac Hucha burials at the hill called Llaver, Musmus, Ucur and Zhio, Cazhalao, the Zhiñang hills, Cumbe, Calvario, Patamarca, etc. (cf. Salazar 2004:63). Many of these sites were later associated with what were described as Catequil *huacas* (see also Verneau and Rivet 1912; Saville 1924; Idrovo 2000a).

recognition of their sanctity (see also Saville 1924; Verneau and Rivet 1912; Salazar 2004). However, the most important of the Inca constructions in this region of the cordillera was the imperial administrative center of Tomebamba (Idrovo 2000b).

Spanish chroniclers state the Inca ruler Tupac Yupanqui (reign: c. 1471–1493) conquered the Cañari during their initial expansion into these regions after subjecting the Huancabamba[25] to the south (Salomon 1987; Hocquenghem 1994). Before the arrival of the Spanish, Huayna Capac (1493–1527) continued Inca expansion into northern Ecuador and, in the process, incorporated the cultures of these regions into the empire. Huayna Capac is said by some chroniclers to have been born in Tomebamba and to have commissioned numerous Inca constructions along the Camino Real, including various rest stops or *tambos* (or *tampus*), which were essentially inns and storehouses, *ushnus*, as well as the stone temples at Tomebamba, Ingaprica, and Mulluturu (see, e.g., Hyslop 1990; Idrovo 2000b; Covey 2006).[26] According to chroniclers, the Orejones of Cuzco—that is, the most learned and noble men in the kingdom—stated that the ruler Inca Tupac Yupanqui was the founder of Tomebamba and that because of the warmer climate he "enjoyed being here more than in any other place" in his kingdom (Cieza de León 2005 [1553]:169). They also affirm that while residing in Tomebamba Huayna Capac first heard of the first arrival of the Spaniards along the Tumbez coast from one of his messengers, "when Don Francisco Pizarro reached the coast in the ship with thirteen companions, who were the first discoverers of Peru; and that he said that, after his days, a strange people would rule the land, like those who had arrived in the ship" (ibid.:169). After conquering the Cañari, Tupac Yupanqui ordered the construction of a temple to Viracocha called Pumapungo or "Door of the Puma," and the surrounding legend and folklore emphasize its once golden temples and incredible riches, its objects of precious metal and jewels, suggesting that this temple symbolized a place of emergence (**Figure** 17; see color plate section). Significantly, Pumapungo is demarcated by a series of artificial-stone-faced terraces—architectural modifications and characteristics later described by Augustinian clerics as indicative of lightning-related *huacas* associated with Catequil, the cult to lightning (San Pedro 1990 [1560]; Albornoz 1989 [1584]; Molina 1959 [1575]; Idrovo 2000b). Cieza de León

---

[25] Scholars in the biological and earth sciences refer to this area of the Andes cordillera as the "Huancabamba Depression" because these latitudes represent the lowest altitudinal conditions along the cordillera. Consequently, southern Ecuador and northernmost Peru have highly diverse ecologies, with a high incidence of endemism of both flora and fauna (Acosta-Solis 1970; Gentry 1986). This part of the cordillera is renowned by locals and scholars alike for it its endemic orchid and hummingbird species (Acosta-Solis 1970; Ridgely and Greenfield 2001).

[26] Mulluturu is an Inca *Hanan ushnu* northwest of Cuenca that was commissioned by Huayna Capac and located beside the Royal Road or Camino Real on a remote mountain peak in the Paredones of Molleturo (Sarmiento 1907 [1572]; Cieza de León 1998 [1553]; Albornoz 1989 [1584]).

(2005 [1553]:165) states, "The famous buildings of Tumebamba are in the province of Cañaris, and they were among the richest and most splendid in the whole kingdom of Peru." The Cañari originally called Cuenca by the name Guapondeleg, and their occupation of this region extends back to c. 500 CE (Idrovo 2000b). Consequently, there were many regional *huacas*, many of which were natural springs, lagoons, and ponds often named in local origin myths as places of emergence into the present world cycle or creation. During Inca expansion they recognized the sanctity of some of these *huacas* through the construction of temples to Inti the sun, as they did at Ingapirca and Tomebamba. With reference to the Temple of the Sun Huayna Capac had constructed at Tomebamba, Cieza de León (2005 [1553]) observed that:

> the temple of the sun is built of stones very cunningly wrought, some of them being very large, coarse, and black, and others resembling jasper. Some of the Indians pretend that most of the stones of which these buildings and the temple of the sun are built, have been brought from the great city of Cuzco... by means of strong ropes. (Ibid.:163)

The Inca did not suppress the Cañari but, rather, absorbed their impressive achievements in astronomy and agriculture (Fresco 1983; Idrovo 2000b). As was customary, the Incas used *mitá*, traditional forms of tribute and reciprocity, to administer their empire and incorporated subject populations in order to maintain religious, political, and economic ties to the Cañari (Murra 1946, 1975a [1972], 1985a, 1985b).

After their conquest of the Cañari, the Inca renamed Guapondeleg after the sacred center at Tomebamba. Tomebamba also served as an administrative center for the northern regions of the Inca Empire and was once said to have held great riches and splendor (Molina 1959 [1575]).[27] When the Spaniards arrived at the legendary center of Tomebamba, all that remained were ruins (Cieza de León 2005 [1553]:162–169). Cieza de León (2005 [1553]) was nevertheless astounded:

> Within the buildings there were several bunches of golden straw, and sheep, lambs, birds, and many other things were sculptured on the walls. Besides all this, they say that there were enormous sums in gold, preserved in jars and vases, and many rich vestments, adorned with

[27] Accounts by the Spanish conquistadores suggest that Tomebamba was destroyed by the Cañari after the death of Atahualpa in Cajamarca on July 26, 1533 (see Staller 2009:456–462). The last Inca emperor, Atahualpa was killed the day after the feast day of Santiago. The Cañari sided with Huascar against Atahualpa in the Inca civil war (Cieza de León 2005 [1553]:166; Salomon 1987; Covey and Elson 2007). After having been abandoned by the Cañari and then the Incas, Tomebamba remained sparsely populated until the 1550s (Covey and Elson 2007; see, e.g., Cieza de León 2005 [1553]).

> silver work and beads. In short, I am unable to describe the magnificence of these royal palaces of the Yncas. (Ibid.:166)[28]

Tomebamba also included storehouses with great quantities of cloth and a palace of cut stone to house more than two hundred princesses, whom the Spaniards called virgins of the sun or *mamaconas*, weavers dedicated to the service of the Sapa Inca, who was perceived to be a living manifestation of the sun (ibid.). According to chroniclers, these women "were very beautiful, and natives of Canaris," and before the Spaniards came to these regions the province was governed by a chief superintendent of the Inca, who resided at Tomebamba and had charge of the temple and the many buildings behind the Inca palaces (Cieza de León 2005 [1553]:166; Idrovo 2000b:16–19). Some of these Inca buildings were said to have served as lodging for Inca soldiers, and accounts emphasize that the storehouses were kept full (Cieza de León 2005 [1553]:163; see also Uhle 1923; Idrovo 2000b).

Spanish colonial rule and the conversion of local populations from Illapa to Santiago was particularly important to local clerics because of the history of animosity and conflict between the Inca and Andean populations on the opposite sides of the Andes in southern highland Ecuador (Sarmiento 1907 [1572]; Albornoz 1989 [c. 1584]; Guaman Poma 1980 [1583–1615]; Salomon 1987; Covey and Elson 2007). These early Spanish accounts provide clear evidence of a symbolic link between Illapa and precious metals through Illapa's representation and ritual associations to objects made of precious metal such as gold and silver, and in many cases to sites that were also *huacas* associated with the metallurgical arts (see also Idrovo 2000a, 2000b). Such ties may explain why Illapa was represented in Cuzco as a faceless golden idol that was kept in a special sanctuary in the Coricancha, and also why many of the *huacas* in these regions and in highland Bolivia around Potosí that were associated with the ceramic and metallurgical arts were initially associated with lightning veneration, and in later periods with veneration to Santa Barbara, Santiago, and the Immaculate Conception.

The importance of converting indigenous populations through religious suppression and intolerance of ritual and religious veneration surrounding Illapa is apparent in the city of Cuenca. It was from Cuenca that the Augustinian clerics and the colonial government operated their campaigns to eradicate such idolatry and the associated rituals and rites. An early colonial concern regarding conversion through the cult of Santiago is clearly represented by the historical and cultural importance of the Cathedral of the Immaculate Conception (**Figure** 18); see color plate section).

---

[28] This description suggests that Tomebamba was regarded by the Spanish conquistadores as one of the so-called Cities of Gold often mentioned in legend and folklore surrounding El Dorado.

The Cathedral of the Immaculate Conception has long been considered the spiritual and cultural heart of the city of Cuenca. Its three domes symbolize the colonial history and reflect the importance of the concept of an immaculate conception, first perpetuated by the Crown and the Order of Santiago and its associated rituals and rites, to this region of the cordillera. European royalty appear to have taken notice: in 1646, King John IV proclaimed the Lady of the Immaculate Conception patron saint of Portugal. In 1760, Pope Clement XIII granted a request by the Spanish Crown to proclaim the Immaculate Conception patroness of Spain (Cardinale 1983). The feast of the Immaculate Conception is on December 8, as proclaimed by Pope Pius IX, and in 1845 became an obligatory day celebrating that Mary, mother of God, was born without original sin (Madariaga 1947; Cardinale 1983). Coincidentally, the feast day of Santa Barbara, who also has a close association to lightning in some regions of the Andes, is December 4, at the beginning of the rainy season and during the annual period when thunderstorms are most intense. Some Andean communities declared December 4 an obligatory day of rest out of fear that Santa Barbara would send a bolt crashing down to punish those who dared work their fincas (Costas Arguedas 1967/2:211). Significantly, Illapa was also venerated each year in early December during the Inca festival of Capac Raymi (Cobo 1990 [1653]; Zuidema 1980, 1992, 2011). In Cuzco, three statues of the Illapa trinity were brought out of the Coricancha along with the three idols representing Inti the sun during the Capac Raymi (Acosta 1962 [1589]:268).[29] Capac Raymi rituals and rites were focused on penitence and blood sacrifice (Cobo 1990 [1653]; Zuidema 1992, 2011).

An association of Illapa and the Catequil cult to lunar cycles and eclipses may be related to the December solstice, which has celestial associations to dark cloud constellations and symbolic reference in the natural world to felines, serpents, and foxes. The exploitation and reproductive cycles of terrestrial foxes are governed by lunar cycles, and the word *atoq* (fox) has a free association to *wañumarka* or "storehouse of the dead" and serves as the name of the principal *apu* in Misminay, Peru (Urton 1981:70). Thus such animal familiars are also linked symbolically to celestial phenomenon as well as mountain summits.

Fox- and serpent-shaped dark constellations in the center of the Milky Way are particularly associated with certain points in the rainy season, especially December through February. Inca festivals held in early December sometimes referred to dark cloud constellations such as Machacuay "the serpent," which was visible in the night sky during this time in the annual cycle. Thus there are connections among dark constellations, serpents, and water.

---

[29] Triadic associations and classifications are common to Illapa veneration; later, in postcolonial times, they came to have an association with the Holy Trinity of Christian dogma.

Since meteorological serpents (rainbows/*amarus*) appear only during the rainy part of the year, they exhibit a seasonal activity cycle similar to that of the most intense lightning storms—in other words, associations by observational logic.

During the Colonial period Spanish clerics, particularly the Augustinian and Dominican orders, recognized early on the pre-Conquest basis for lightning veneration and were soon intensely focused on replacing its surrounding rituals, rites, and beliefs (e.g., twin births) with veneration to the cult of Santiago. Consequently, colonial authorities did not seriously attempt a campaign to root out its pagan roots until the seventeenth century, when all forms of native animism were highly suppressed and punished by the various colonial governments and the Church. As part of that effort, the clerics forbid subject populations to bestow the name Liviac or even Santiago to indigenous babies born during a lightning storm (Arriaga 1968 [1621]:215; Gades 1983:779).

Some local colonial *huacas* or shrines (*villcas*) in the Andean countryside had strong lightning associations (Duvoils 1967:18–19). Chordeleg was such a site. Situated less than fifty kilometers east of Cuenca, it was an indigenous town known for its ceramicists, weavers, and embroiderers during the Conquest. The surrounding hilltops included many ancient Cañari metal crafting centers and pre-Columbian cemeteries. The strong Inca presence in this region may also have been related to gaining direct access to status-related objects of gold, silver, and cloth (**Figures** 19a–b). The Inca built structures at and architecturally modified many of the *huacas* in these regions of the Andes, particularly those near the Camino Real, in recognition of their religious and mythological importance to their subject populations (Uhle 1922, 1923; Gonzalez Suárez 1878, 1892; Espinosa Soriano 1988a). Andean societies of this region were renowned throughout the Andes for their ceramics and metallurgical crafts, and these technologies have ancient origins in this part of the cordillera (Salazar 1998, 2004; Salomon 1987; Staller 2003. 2007). Archaeologists identified a stone corridor in the form of a snake near Chordeleg built in pre-Inca times, again suggesting possible lightning associations. Another important *huaca* is Llevar, located northwest of Chordeleg (Salazar 2004:55). Llevar is a terraced hilltop in the Paute-Gualeco Valley with mythological associations to the Cañari, who believed that their mythological ancestors were serpents that emerged into the present world cycle from lagoons and mountain lakes (Salazar 2004:54–55, 66–67; González Suárez 1878). Its architectural modifications are similar to hilltop and clifftop sites identified by Max Uhle (1922) in his archaeological survey of this region. Another major metallurgical site with a pre-Columbian necropolis is Sigsig (Saville 1924; Salomon 1987; Verneau and Rivet 1912). Sigsig is renowned for the metal artifacts looted from that area and is believed by some scholars to be the original home of the famous Sol de Oro, a gold mask believed to have been looted from a tomb in Patecte, near Chordeleg, that has become an icon, the logo representing the Museo Antropológico del Banco Central del Ecuador (**Figure** 20a; see color plate section). Museum officials told me that the gold mask was indeed taken from an elite tomb at the ancient necropolis of Chuncari near Sigsig. Jagged serpents emanate like bolts of lightning from the human face in the center of the mask, their undulating bodies similar to the double jagged lines used to represent Illapa in the Inca cosmology by Santa Cruz Pachacuti Yamqui and suggest a symbolic reference to Illapa. Such a double-jagged-line motif appears in Formative period pottery from the hilltop *huaca* of Cerro Narrío (**Figure** 20b; see color plate section).

Lightning *huacas* have also been reported farther to the south in the Bolivian altiplano, another important mining region. A lightning *huaca* called Collana is situated on a hilltop high in the altiplano of Bolivia in the lofty interfluves of La Paz (Lunardi 1946). Lightning strikes at this altitude hit the site with legendary frequency, explaining why Collana was such an important pre-Conquest *huaca* to Catequil, the cult of Illapa. In fact, the *huaca* was so important that the Inca constructed a fortress (*pucará*) beside it. In the sixteenth century,

FIGURE 19A French archaeologist Paul Rivet beside Inca ruins c. 1912 at Chordeleg, Azuay Province, Ecuador. Chordeleg was an important prehispanic *huaca* with an ancient necropolis. (Collection of G. Landívar; from Salazar et al. 2004: 58)

FIGURE 19B The German archaeologist Max Uhle standing beside Inca ruins in 1922, at the temple center of Ingapirca, Cañar Province, Ecuador. Uhle discovered the cultural horizon in southern highland Ecuador called Chaullabamba. His excavations at Pachacamac in coastal Peru identified the Middle Horizon with earlier Wari expansion. (Collection of Matthias Abram; from Salazar et al. 2004: 22)

the Spaniards directed construction of an indigenous village and named it Santiago de Collana in an attempt to convert local Andeans (Gade 1983:779). Many *huacas*, including large numbers that were dedicated to Illapa as a harbinger of rain, were destroyed in this period in the Ayacucho area alone (Duvoils 1967:19). However, numerous specialized lightning *huacas* survived the campaign, and new ones subsequently emerged long after the Colonial period (Middendorf 1895/III:419; Bastian 1978:69).

Shells of Lightning: Fertility, Rebirth, and Death

One of the important pre-Columbian symbols of veneration to Illapa, the thorny oyster, continued to play a role in the rituals surrounding lightning veneration in post-Conquest times despite considerable colonial resistance to such practices. Thorny oyster (*Spondylus* spp.) or *mullu* was central to early long-distance interactions traded as crafted status-related objects, symbolically represented on pottery and in iconography from the very early periods of

Andean prehistory (Paulsen 1974). *Spondylus* shell was highly valued for its red color, and worked beads were used as a form of currency just before the Conquest period.[30] Significantly, *Spondylus* shell ornaments and ritual offerings were commonly associated with Illapa and its related *huacas* and *ushnus*, an association that appears to be very ancient (Paulsen 1974; Staller 2007, 2010b; Temme 2000). There may be cultural reasons for such associations: in the Inca cosmogony, Illapa was perceived as brother to Earth Mother, or Pachamama (Sullivan 1985). In fact, *Spondylus* has a multifaceted role in Andean cosmology as well through female associations as a symbol of sexuality, fertility (both agricultural and human), and rain (Sharon 2001). *Spondylus* shell objects were commonly left as ritual offerings at *huacas*, especially subterranean springs, and places where rivers split as well as mountain passes and summits. Indigenous populations currently use the Quechua term *mullu* as synonymous with *Spondylus* shell (Murra 1975b; Hocquenghem 1993b). *Spondylus* spread along the Peruvian coast beginning in the Initial period and is particularly apparent along the north coast with Moche and Chimu culture (Pillsbury 1994; Sharon 2001; Donnan 2004). Such associations provide a basis for understanding its early importance to early long-distance interaction in the highlands and along the coast (see, e.g., Paulsen 1974; Murra 1975b; Salomon 1987; Hocquenghem 1993a, 1994; Salazar 1998; Cordy-Collins 2001a, 2001b).

Thorny oysters are symbolically associated with female sex through concepts and beliefs surrounding *vagina dentata*, and thus have a logical connection to childbirth, to twins, and indirectly to the fertility of the earth (Paulsen 1974; Pillsbury 1996; Cordy-Collins 2001a, 2001b). Cultural associations between *Spondylus* and lunar cycles and eclipses provide a basis for understanding symbolic lunar associations to Catequilla and the superstition involving lunar eclipses mentioned repeatedly in the Augustinian chronicles. The *Strombus* conch was valued in traditional culture as a horn or *pututu* and noted for its complex eyes, which are situated at the ends of long, stout stalks (Keen 1971:420). *Strombus* conch is a terrestrial species adapted to intertidal zones and grazing on fine, filamentous algae. *Strombus* spp. have the same southerly range as thorny oyster and are found on sand flats, in muddy lagoons, and offshore to depths of forty-five meters (Keen 1971:420–421). This indicates that all such mollusks found in ancient sites along the Peruvian and Chilean coast were transported from southernmost Ecuador and far northern Peru and regions to the north (**Figures** 21a–b; see color plate section). The fact that it remained closely associated with Catequil ritual practices even in the Colonial period suggests that such interaction

[30] Thorny oyster (*Spondylus* spp.) is a marine shell species that ranges along the eastern Pacific from the Gulf of California to southern coastal Ecuador. Spaniards in Central and South America later used this thick-shelled species to make lime for cement (Keen 1971:96).

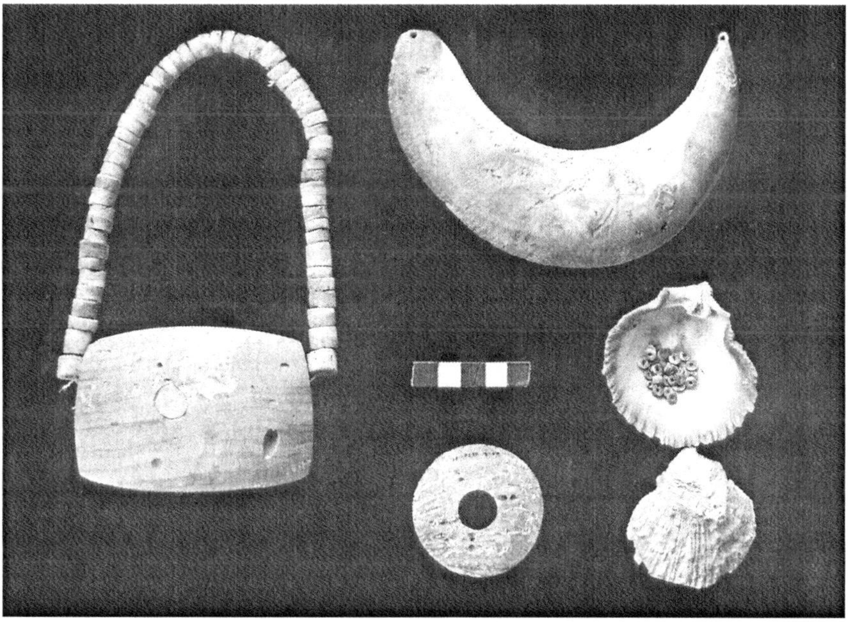

FIGURE 21C Worked and unmodified thorny oyster (*Spondylus* spp.) as well as metal artifacts were commonly left as offerings at lightning *huacas* in the Late Horizon period Capac Hucha burials. *Spondylus* necklaces and bead manufacturing sites extend back to the Formative periods on Puná Island in the Gulf of Guayaquil, where numerous workshops were uncovered by archaeologists. (Courtesy of Museo Antropológico del Banco Central del Ecuador; photo by John E. Staller)

was of great antiquity and continued to have ritual associations to Illapa after the Conquest. Both *Strombus* and *Spondylus* appear in iconography and pottery extending back to the Initial and Early Horizon periods in Peru and as status-object offerings in earlier Formative and Preceramic period contexts. *Spondylus* is particularly prominent in Chimú iconography, where the moon rather than the sun was the primary focus of religious veneration (**Figure** 21c).

It is highly significant, in regard to the importance of thorny oyster and *Strombus* conch to ancient Andean religions and religious iconography, that pecten (*Argopecten circularis* Sowerby), the calico scallop, is associated with Santiago almost exclusively, and not to Illapa and its associated rituals and rites (**Figure** 22). The relics of St. James in Santiago de Compostela have over the centuries drawn large pilgrimages on July 25 to celebrate the rituals and rites surrounding the cult of Santiago, in this part of Galicia, Spain (Stokstad 1978; Cardinale 1983). A shrine in the Cathedral of Santiago de Compostela is known in English as the Way of St. James (Camino de Santiago), and Santiago de Compostela is associated with Celtic legends as a sacred place where souls of the dead gathered to follow the sun across the ocean (Stokstad 1978).[31] When the holy relics were declared those of St. James, King Alfonso II of Oviedo had the Cathedral of Santiago de Compostela built on the spot where his remains were believed to have been discovered (Stokstad 1978). The legend also includes numerous events perceived as miraculous. One which particularly united the Spanish was that Santiago de Compostela had enabled the Catholic faithful to maintain their stronghold in northwestern Spain and also helped them defeat the Moors during the various crusades, thus St. James as Santiago Matamoro had close symbolic associations with the birth of Spain and the Holy Roman Empire (Cardinale 1983). The scallop shell is the traditional emblem of St. James and is popular with pilgrims to the apostle's shrine at the Cathedral of Santiago de Compostela. Religious pilgrims on the Way of St. James often wore a scallop shell symbol on their cloths or around their necks. During medieval times, pilgrims traveling to the shrine would be given as much sustenance—wine, beer, oats, barley, and so on—as they could pick up with one scoop on presenting the scallop shell to a church, castle, or abbey (Cardinale 1983; Stokstad 1978). Numerous pilgrimage roads to Santiago de Compostela are marked by the plaques or painting of the shell, which is now synonymous with Santiago. The scallop shell also has links to twin births and fertility, related in part to its shape. In Greco-Roman iconography and imagery, scallop shells symbolized the feminine principle, perhaps through their symbolic associations with the goddesses of love and fertility, Aphrodite and Venus respectively (Cardinale 1983). Another symbolic tie connecting pecten shells to Santiago is related to a legendary miracle involving the recovery from death of a knight "covered in scallops," thus a direct metaphor to life after death. Although the scallop maintains a somewhat close association to Santiago and the cult of St. James in the New World, neither the Augustinians and other religious clerics nor the Inquisitors were as successful in replacing the *mullu* or thorny oyster (*Spondylus* spp.) and the *Strombus* conch (*Strombus galeatus* L.) with the indigenous sea scallop (*Argopectan gibbus* L.) as a symbol of Santiago Illapa in the Andes (Espinosa Soriano 1988a; Topic et al. 2002) (**Figure** 23). Significantly, the Cañari venerated the moon and had a lunar calendar (Espinosa Soriano 1988a). Their circular temples stand in contrast to the rectangular Inca constructions at sites such as the imperial centers of Tomebamba and Ingapirca (see Fresco 1983:Figs. 2, 12; Idrovo 2000b:86–87) (see **Figure** 17).

[31] Santiago de Compostela was a Roman cemetery by the fourth century, which may be in part related to such later Spanish legends and folklore (Fletcher 1984:57–59).

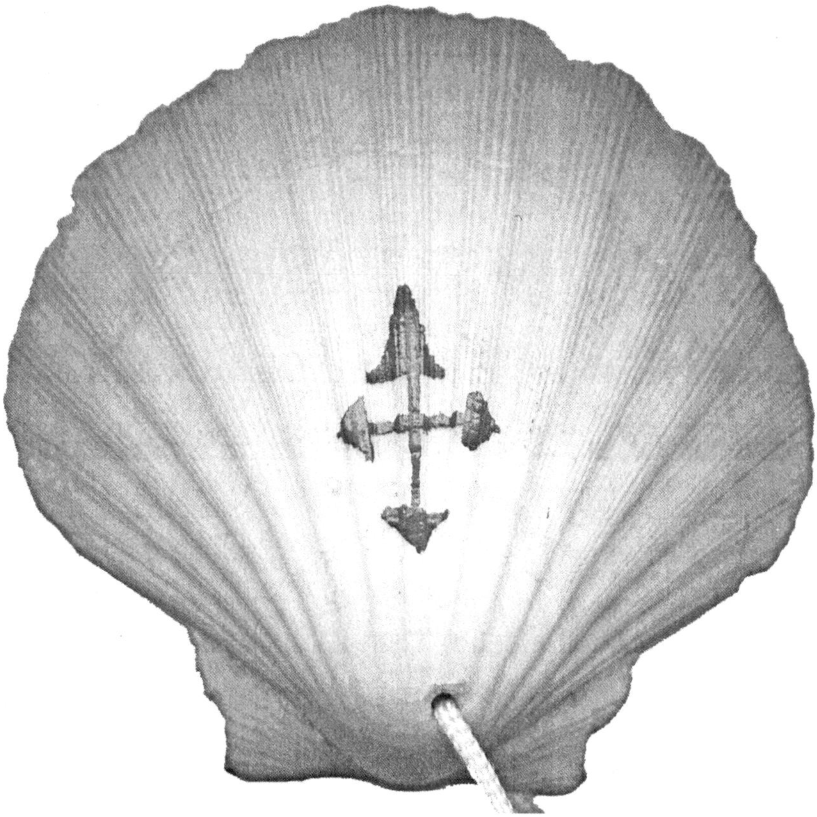

FIGURE 22 Atlantic calico scallop (*Argopecten gibbus* L.) or pecten shell of St. James (Santiago). Paintings and plaques of this shell mark the various roads or Way to St. James taken by pilgrims traveling to the shrine in Santiago de Compostela. Its symbolic associations also extended to the Immaculate Conception.

Catequilla: Lightning *Huacas* and Inca Expansion

In Ecuador and northern Peru, Colonial period lightning veneration was associated with the primary constituents of the Catequil cult (Topic 1992, 1998; Topic et al. 2002). Traders or *mitmaqkuna* or *mindalá* are believed to have brought the religious cult to these regions during Inca expansion, but recent archaeological evidence suggests that archaic forms of lightning veneration extending back to the formative periods may have existed in non-Quechua-speaking regions south of Riobamba (see, e.g., Salazar 1998; Idrovo 2000a; Temme 2000; Staller 2007; Villalba 2011). Coastal societies cultivating maize and with direct access to *Spondylus* spp. and *Strombus galeatus* conch initially spread in the context of early long-distance interaction, based on direct access

FIGURE 23 The South American scallop (*Argopecten circularis* Sowerby) has a range of colors and markings, from pure white to blotched and streaked patterns to solid dark orange and purple, and can be found as far south as Paita, Peru. Colonial governments and the Church were unsuccessful in replacing the *Spondylus/Strombus* dyad with this indigenous species of scallop. (Photo by John E. Staller)

to strategic nonlocal resources and consumables (Paulsen 1974; Murra 1975a [1972]; 1975b). The integration of ceramic technology and spread of domesticates such as maize suggests a widespread shift to an agricultural economy associated with such early interaction (Staller 2007, 2010b; Idrovo 2000a; Temme 2000; Villalba 2011).

The non-Aymara/Quechua-speaking societies of southern highland Ecuador and northernmost highland Peru in part explain the absence of the Catequilla toponyms in these regions. Moreover, many of the *huacas* in these regions are associated with lagoons and mountain summits, presumably places of emergence into the present world cycle, and many of their most important sites were involved in the metallurgical arts (Staller 2007). Consequently, offerings at local *huacas* included status-related objects of precious metal, thus many of the lightning-related sites have been extensively looted since the Colonial periods in these regions. Chroniclers and ethnohistorians maintain Catequil has pre Inca origins in Ecuador related to a pan-Andean thunder and lightning cult (San Pedro 1992 [1560]:174; Espinosa Soriano 1988a, 1988b). Such assertions and accounts are largely consistent with Inca and pan-Andean cosmogony, particularly to intermediate periods of destruction and chaos, and the beginnings of the present creation or world cycle (Pease 1973; Sullivan 1985).

Augustinian friars ordered thousands of lightning *huacas* (Illapa *ushnu*) destroyed, and the most important was the Oracle of Catequil de Huamachuco (Topic 1992; Topic et al. 2002). Located on a high hill with three large cliffs near San José de Porcón, the oracle at Huamachuco was located on a mountaintop near three cliffs called Apocatequil, Mamacatequil, and Piguerao. On Apocatequil there was said to be a stone statue or *yañca* representing the *huaca* believed to make other *huacas* "speak," and only the divine ruler (Sapa Inca) was believed to make *huacas* "speak" (San Pedro 1992 [1560]:177–178).[32] Cristobál de Albornoz (1984 [1584]:210) states that Catequil "was some tall rocks on a plain, and as in the air above the rocks" (cf. Topic et al. 2002:310). The *huaca* has been identified as the southeastern face and summit of Cerro Icchal, which rises to 4,150 meters above sea level (Topic et al. 2002:308–309). Both Juan de Betanzos (1987 [1551]:250) and San Pedro (1992 [1560]:176) state that the Inca ruler Atahualpa destroyed Catequil by fire and that after the stone idol had been destroyed, the priest still continued to venerate the cliffs (San Pedro 1992 [1560]:178). Betanzos (1987 [1551]:250) also states that Atahualpa systematically destroyed the oracle at Huamachuco because of a "response" the Inca ruler received while "consulting" with it. Betanzos also states Atahualpa considered Catequil, like Huascar, his enemy. Atahualpa spent three months at the *huaca* at Huamacucho supervising its destruction (Betanzos 1987 [1551]:251). This occurred during a crucial stage of the Inca civil war with Huascar: while Atahualpa single-mindedly devoted himself to the destruction of Catequil de Huamacucho, Huascar was captured; his faction in Cuzco was massacred at the same time that news arrived of the

[32] The most detailed information on the oracle at Huamachuco comes from Fray Juan de San Pedro in his *Relaciones de Primeros Agustinos*, written in Spain in 1560 or 1561 (Topic et al. 2002). The lightning *huaca* at Huamachuco was a *yañca*, one that could make other *huacas* speak (Rostworowski and Morris 1999:796).

landing of Pizarro and his ships along the coast at Tumbez.[33] Catequil de Huamachuco was clearly considered important by members of the Inca elite, and this importance extended to factional disputes and to communication with the natural world. Cristobál de Molina, "El Almagrista" (1968 [1552]), describes it in the context of the Inca civil war and Spanish *entrada*, stating that Atahualpa destroyed the oracle because it had foreseen the Christian victory. The Jesuit José de Arriaga (1968 [1560]:200) attributes the destruction of the *huaca* of "Catequilla" in Huamachuco to the Inca Huascar, "hijo de Topa Inga," who put fire to the *huaca*, but he reports that the priests rescued the idol and brought it to Cahuana (modern Cabana, Conchucos), where a new shrine was built, and later destroyed Fray Francisco Cano (Topic et al. 2002:307). Arriaga further states that some believed the idol was removed to Tauca (also in Conchucos), where it is hidden still. Juan de San Pedro (1992 [1560]:177–178) and Juan Betanzos (1987 [1551]:250) emphasize that the cliff and the hill itself was burned, while the chronicler priest Pedro Sarmiento (1907 [1572]:176) says the hill was leveled, thus portraying the elemental battle between supernatural protagonists employing fire and water, the essential elements associated with lightning. Their emphasis on the hill confirms that the mountain was the vitalizing force or animating essence of the *huaca*. These chronicler accounts also mention the idol and say the high priest had been decapitated (Topic et al. 2002:312).[34] Lightning was so important to Inca culture that Guaman Poma (1980 [1583–1615]:62, fol. 79) included it in their coat of arms, along with the sun, moon, and Paqariqtambo, the place of their mythological origins (**Figure** 24). The location and placement of the primary sources of Inca cultural identity and religious veneration may be examined with regard to Inca cosmogony, which perceives lightning or Illapa as son of Inti the sun, while the moon (Quilla) had symbolic associations to Pachamama or Earth Mother, who according to Inca creation myths was the "sister" of Illapa. In the Inca coat of arms, Guaman Poma depicted Illapa below the sun and beside Pacariqtambo, the place of their mythological origins. It was from the three caves or windows that the original Inca emerged into a wild landscape (Urton 1990, 1999). Illapa also has close symbolic associations to caves and lunar eclipses.

Among Andean populations during the period of the Spanish conquest, lightning had symbolic and metaphorical references to twin births, physical

[33] The Cañari sided with Huascar in the Inca civil war. Moreover, Cañari elite were in Cuzco when Atahualpa's army attacked, and they suffered many casualties (Salomon 1987; Covey and Elson 2007). The Cañari had a history of conflict with the Inca in resistance to and campaigns against Tupac Inca Yupanqui and his son Huayna Capac (Sarmiento 1907 [1572]; San Pedro 1992 [1560]).

[34] It is noteworthy to point out that decapitation sacrifice has been documented in both the pre-Columbian coast and highlands, although it has not been directly connected to lightning veneration (see, e.g., Chavez 1992; Cordy-Collins 2001a; Arnold and Hastdorf 2008).

FIGURE 24 Inca coat of arms, "Armas Propias." Guaman Poma de Ayala includes Inti the sun, Coya the moon, Illapa the lightning, and Pacariqtambo, site of their mythological origins. Above and below the symbol of lightning he wrote, "Choqui Ylla Uillca" (*el noble del rayo, o de oro*) the noble bolt or gold, in reference to the golden idol without a face that stood in the Coricancha or Temple of the Sun in imperial capital of Cuzco. (From Guaman Poma 1980 [1583–1615]: 62, fol. 79)

deformations such as harelips or cleft palates, the principles of renewal and rebirth, solar cycles, and lunar eclipses (Topic et al. 2002; San Pedro 1992 [1560]). The beliefs, rites, and rituals associated with lightning during the early Colonial period were similar throughout the most of the Andes and show an association to regions involved in crafting of objects made of precious metals. These associations to the metallurgical arts and metalworking sites are particularly apparent in the southern Ecuadorian highland and the highlands of northern Peru, as mentioned above and also in the Bolivian altiplano.

Lightning, Metal, and Death in the High Andes

Lightning also plays a central role in the transcendence between life and death and therefore has direct relevance to Judeo-Christian religions, where resurrection from death and immortality are a central characteristic of the human soul as well as gods and divine beings. Religious syncretism between indigenous and European introduced religious customs and beliefs regarding the transcendence of life and death were directly relevant to lightning veneration. Accounts indicate that people were sometimes reported to have come back from the dead after being struck by a lightning bolt. In fact, lightning mortality during the Colonial period was believed to be a form of divine retribution, narrow escapes, and quick recoveries attributed to the mercy of God. In 1643, a flash hit a devout man, melting his gold buttons but leaving him essentially unharmed. His narrow escape was attributed to his veneration of Our Lady of Candelaria (Gade 1983:775). In 1669, an indigenous Andean father of five children was struck and believed killed; his wife prayed, and soon thereafter he "came back to life" (Gade 1983:775). In a different incident, a lightning bolt crashed into a bedroom of a house, leaving two sisters unconscious. Assuming they had perished, their mother went to pray for their souls, only to find them alive when she returned from the church. Such recoveries from lightning bolts were attributed to the mercy of God. More recent medical research has documented spontaneous recovery from lightning jolts that paralyze the nerves and momentarily stop the heart (Taussig 1968).

La Plata, Bolivia, was well known for its violent lightning and thunder during the Colonial period (Ramirez del Aguila 1978:29; Cobo 1990 [1653]). The mountain summits that loom above the city, particularly Sicasica and Churuquella, are known to attract lightning bolts, which commonly crash around and below them. In 1610, during a single storm, five different strokes hit the town of Chuquisaca and electrocuted five people. In another tempest, twelve bolts struck; they caused no deaths, but much destruction and injury (Cobo 1990 [1653]:73). After the Recoleta monastery was built in 1600 on a hill above the city, lightning strikes were said to have decreased (Garcia Quintanilla

1963/III:109–110). However, in 1649 several people were killed by lightning on the feast of the Immaculate Conception, including a friar in his convent (Hurtado de Mendoza 1976 [1560]:139). In 1650, large crosses were erected around the city and on the summits of two peaks, because they were believed to provide protection from violent storms (Ramirez del Aguila 1978:29; Gade 1983:774).[35] This is an example of syncretism regarding lightning veneration, as the *cruceros* or crosses were closely associated with Illapa and mountain summits in Andean culture (Cruz 2009).

In the last part of their expedition into the altiplano the Spaniards crossed the Río Pilcomayo and climbed into the silver-mining region of Qaraqara, where they passed the hot springs baths at Chaqui, where the ruler Huayna Capac liked to bathe. A previous ruler, Inca Pachacuti Yupanqui, is said to have brought hammered sheets of refined silver from Porco in order to wrap cut stones associated with the Temple of the Sun in Cuzco in the mid-fifteenth century (Platt and Quisbert 2007:115). From this location the Spaniards reached the silver mountain of Porco, a major lightning *huaca*, in this case an *apu* (mountain lord) venerated throughout the Inca Empire (Cruz 2009:56–57, Fig. 1). Chroniclers mention that other important *huacas* in the region of Potosi and Chuquisaca included Quiquijana and Poder de Dios (Cruz 2009:Fig. 1). The lightning *huaca* of Porco is said to have been symbolically represented by three stones and a lump of pure silver ore, a metaphorical reference to the various triadic classifications associated with Illapa and the *mama* or concentrated fertility (silver ore) of the *huaca* (Platt and Quisbert 2007:115). Many of the mountains in this and surrounding regions have been found to include interments on their summits (Duvois 1976). The *huaca* on the summit of Porco includes a sanctuary consisting of two platforms demarcated by artificial terraces or *murros* constructed of fieldstone (Cruz 2009:59, Figs. 2–3). Cruz goes on to mention several other *apu* in the Cordillera de Los Frailes that included Capac Hucha burials and/or architectural modifications on their summits: Llullaillaco, Chuscha, Picchu Picchu, El Toro, and El Pomo, to name a few (see also Duvoils 1976; Reinhard 1983; Bouysse-Cassagne 1993; Ceruti 1999, 2004; Reinhard and Ceruti 2010). The Capac Hucha burials on the summits of these *huacas* reaffirm, on the basis of the evidence presented in this study, a direct or indirect association to Illapa (Reinhard and Ceruti 2010). Chroniclers mention that when native Andeans would pluck potatoes from the ground around Porco, specks of silver would be sticking to the roots (Platt and Quisbert 2007:115).

---

[35] During the 1980s, the Bolivian government had numerous lightning poles built on the tallest mountain summits surrounding La Plata and Potosí, resulting in significantly lower casualties associated with lightning and the related strikes.

The entrance of the Porco mine was guarded by what the Spaniards described as a *chamán* or shaman who administered "sacraments" to those who entered (Platt and Quisbert 2007:115). These "sacraments" consisted of psychotropic plants, presumably hallucinogenics, administered so miners would experience a hierophany that provided spiritual "communication" with the spiritual essence who was the patron of silver, lightning (*qhaqya* [Aymara]), guiding them to the veins of ore (Platt and Quisbert 2007:115). Although Porco was an important *huaca* and source of Inca silver, the richest mountains were located around Potosi, a few leagues distant from the Porco. Rich Mountain or Cerro Rica, the silver mountain of Potosi, was of particular importance to the conquistadores and the Spanish Crown. According to Cieza de León:

> The mines of Porco, and others in this kingdom, have been open since the time of the Yncas, when the veins whence they extract the metal were discovered; but those which they have found in the hill of Potosi concerning which I now desire to write were never worked until the year 1546. A Spaniard named Villaroel was searching for vein of metal with some Indians, when he came upon this wealth in a high hill, being the most beautiful and best situated in all that district. As the Indians call all hills and lofty eminences Potosi, it retained that name. Although Gonzalo Pizarro was then waging war against the viceroy, and the whole kingdom was troubled with this rebellion, the skirts of the hill were soon peopled, and many large houses were built. The Spaniards made their principal settlement in this place, the court of justice was removed to it, and the town of Plata was almost deserted. They discovered five very rich veins on the upper part of the hill, called the "rich vein." . . . This wealth became so famous, that Indians came from all parts to extract silver from the hill. The climate is cold, and there are no inhabited places in the vicinity. When the Spaniards had taken possession, they began to extract the silver, and he who had a mine gave each Indian who entered it a marc, or, if he was very rich, two marcs every week. So many people came to work the mines, that the place appeared like a great city. That the greatness of these mines may be known, I will say what I saw in the year of our Lord 1549 in this place, when the licentiate Polo was corregidor of the town of Plata for his Majesty. Every Saturday the metal was melted down in his house, and of the royal fifths there came to his Majesty thirty thousand or twenty-five thousand pesos, and sometimes forty thousand. And while extracting such immense wealth, that the fifth of the silver, which belonged to his Majesty, came to more than one hundred and twenty thousand castellanos every month, they said there was little silver, and that the mines were not well worked. Yet this metal which

> was brought to be melted, was only what belonged to the Christians, and not even all that, for a great deal taken in pure bits and carried off; and it may be believed that the Indians took a great deal to their own homes. (Cieza de León 2005 [1553]:386–388)

Located in the Bolivian highlands at 4,090 meters (13,420 feet) above sea level, Potosí has a long oral and written tradition surrounding lightning. Potosi was the location of the Spanish colonial mint, associated in this case with Cerro Rico (Cerro de Potosí), one of the richest silver ore deposits in all of the Andes (Matienzo 1967 [1567]; Arzáns de Orsúa y Vela 1965).[36] Acosta (1962 [1589]) stated that in his time there were four principal veins of silver on the hill of Potosí; the Spaniards called these veins La Rica, Centeno, Estajio (tin), and Mendieta. In La Rica there were seventy-eight mines, many of which were very deep. To remedy the evils caused by their great depth, miners made horizontal excavations called *socabones* in the sides of the hill, extending them until they met the vein. The precious metal of the hill at Potosí was discovered by an Indian named Hualpa, a native of Chumbivilica near Cuzco (Arzáns de Orsúa y Vela 1965). He was climbing up a steep part of the hill in pursuit of deer and helping his ascent by catching hold of the quinoa shrubs (*Chenopodium quinoa* Willd.) that grow there. One of the shrubs came up by the roots and disclosed a quantity of native silver, which was the commencement of the vein called La Rica (Cook 1981:236–237, 1998:127). According to colonial documents, between 1556 to 1783 over 41,000 metric tons of pure silver were mined, and from this total, 8,200 metric tons or 20 percent went directly to the Spanish Crown. Numerous scholars have documented on the basis of such colonial records or *relaciones* that untold numbers of Andean peasants (*yanacunas*) died in forced labor working gold and silver mines in the Colonial period (Cook 1981:237, 1998:169–170; see also Matienzo 1967 [1567]; Acosta 1962 [1589]).

> From the year 1548 to 1551 the royal fifths were valued at more than three millions of ducats, which is more than the Spaniards got from Atahualpa, and more than was found in city of Cuzco, when it was first occupied.... The Indians go to the heights with the ores to extract the silver, and they call the moulds Guayras.[37] In the night there are so many of them on all parts of the hill that it looks like an illumination. When the wind is fresh they extract much silver, but when there is no wind they cannot by any means extract silver; so that, as the wind is

[36] The gold mines of Tipuani, east of the Bolivian cordillera, are believed to be the richest in South America (Cieza de León 2005 [1553]: 388 n.; see also Lechtman 1976).

[37] *Huayra* is "wind" or "air" in Quichua. In this example the term refers to small ovens made of perforated clay pots in which the metal was melted (Cieza de León 2005 [1553]:388).

> profitable in the sea for navigating, it is so here for extracting silver. As the Indians have no overseers when they carry the metal up to the heights, it must be supposed that they have enriched themselves, and taken much silver to their own homes. This is the reason that Indians have come from all parts of the kingdom to this settlement of Potosi, to take advantage of the great opportunities offered for enriching themselves. (Cieza de León 2005 [1553]:388–389)

Consequently, there is a large body of ethnohistoric documents on this region extending back to the very beginnings of the Conquest. This is particularly the case with regard to colonial records or *relaciones* regarding land claims, tax, census information, and so on, between colonial officials and local indigenous societies. Cerro Rico dominates the surrounding landscape of Potosí, its summit standing 4,824 meters above sea level, high in the Bolivian altiplano (**Figure** 25). In 1563, Potosí was solely a town, but nevertheless was granted a coat of arms by Philip II (Matienzo 1967 [1567]; Assadourian 1989 [1568–1571]; Cook 1981, 1998). In 1572, Don Francisco de Toledo the Viceroy of Peru went to this great seat of mining wealth and established regulations and amalgamation for its government (Platt and Quisbert 2007:119–120). Toledo introduced the use of quicksilver, and also regulated and legalized the atrocious system of *mitá*, or forced labor of indigenous Andeans in the mines (Assadourian 1989 [1568–1571]; Matienzo 1967 [1567]). The 1572 census indicated a total of 1,677,697 men between the ages of eighteen and fifty liable for service, and this total was derived from some 614 *ayllus* (Cieza de León 2005 [1553]:388 n.). In 1611, the population included some 160,000 inhabitants in Potosí, of which 76,000 were indigenous Andeans, 3,000 Spaniards, 35,000 Creoles and 40,000 Europeans, and 6,000 Negroes and mulattoes (Cieza de León 2005 [1553]:388–389 n.; Cook 1998). Since the town was the location of the Spanish mint, much gold and silver from all over the Andes was brought there to be melted down or traded for money. Cieza de León points out that

> in all parts of this kingdom of Peru we who have traveled over it know that there are great fairs or markets, where the natives make their bargains. Among these the greatest and richest was formerly in the city of Cuzco, for even in the time of the Spaniards its greatness was caused by the gold which was bought and sold there, and by the other things of all kinds that were sent into the city. But this market or fair at Cuzco did not equal the superb one at Potosi, where the traffic was so great that, among the Indians alone, without including Christians, twenty-five or thirty thousand golden pesos exchanged hands daily. This is wonderful, and I believe that no fair in the world can be compared to it. I saw this fair several times, and it is held in a plain near the town. In one place there were *ustos* [bags] of coca, the most valuable prod-

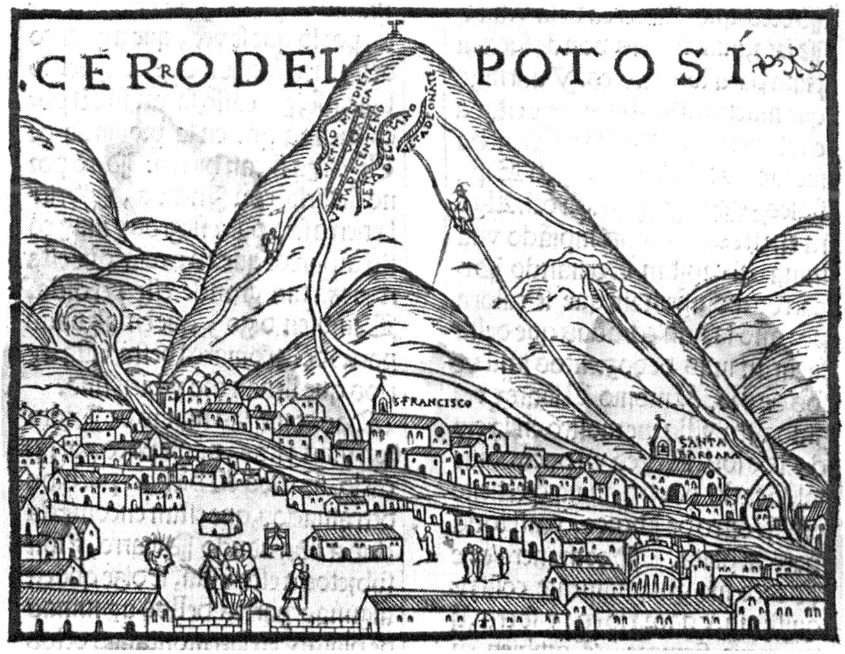

FIGURE 25 Cerro de Potosí (Cerro Rico) in Potosí, Bolivia, as depicted by the conquistador Cieza de León in the first part of his 1553 account of the discovery and conquest of Peru. Founded in 1546, Potosí is situated at 4,090 meters above sea level and is one of the highest Andean cities. A colonial mint was in Potosí, since Cerro de Potosí (Cerro Rico) had some of the richest silver deposits in the New World. Between 1556 to 1783 over 41,000 metric tons of pure silver were mined from this mountain alone. (From Cieza de León 1998 [1553]: 395)

> uct in these parts. In another place there were bales of cloth and fine rich shirtings. There were heaps of maize, dried potatoes, and other provisions there great quantities of the best meat in the country. This fair continued from early morning until dusk; and as these Indians got silver every day, and are fond of eating and treating, especially those who have intercourse with Spaniards, they all spent what they got, so that people assembled from all parts with provisions and other necessaries for their support. Many Spaniards became rich in this settlement of Potosi by merely employing two or three Indian women to traffic in this fair. Great numbers of Yanacuna, who are free Indians with the right of serving whom they please, flocked. (Cieza de León 2005 [1553]:390–391)

The indigenous population in this region experienced a long dark history of oppression relating to the mining of precious metals like silver, as well as a strong colonial presence in Potosí (Assadourian 1989 [1568–1571];

see also Matienzo 1967 [1567]; Arzáns de Orsúa y Vela 1965; Cook 1998; Covey and Elson 2007). Colonial governments in this part of the sierra enslaved tens of thousands of indigenous populations to work the mines, and this had consequences in Andean minds as forms of divine retribution by lightning occurred all around them (see, e.g., Matienzo 1967 [1567]). Churches were struck regularly, and electrical storms and hail would ignite the city, destroying thatch and tile roofs and striking down all those in its path. Colonial documents, primarily interviews with colonial officials and clerics, indicate that in 1597 more than twenty bolts struck dwellings and churches—some even fell into the plazas and streets—killing almost thirty people (Gade 1983:774). In 1650, lightning bolts from another great storm dispatched eleven people in scattered parts of Potosí (Hurtado de Mendoza 1976 [1560]:549; Gade 1983:774–775). In 1710, ball lightning struck down ten people in a span of only fifteen minutes. One of the incidents describes ball lightning flashes entering an upstairs room and killing one person, and presumably leaving the other four alive but unable to speak. The luminous sphere then moved to the ground floor, where it struck down three Andeans and injured five others as they were seeking shelter from the rain. The bishop of Potosí interpreted the lightning deaths as an expression of heavenly wrath toward the sinful Andean inhabitants of Potosí who had accepted "profanity of dress," including a preference for bright colors characteristic of Andean cloths. Expiatory novenas were scheduled (Gade 1983:774). Andeans still continued to perceive Illapa as malevolent protector, rationalizing the destruction set upon the Spanish as vindication for their oppression and intolerance of indigenous religious and spiritual beliefs. The degree to which lightning and later conversion to the cult to Santiago, veneration of the Immaculate Conception etc., affected the folklore and legend of those areas of the Andes, southern highland Ecuador, northern highland Peru, and the Bolivian altiplano supports various lines of evidence indicating that Illapa had strong symbolic and metaphorical associations to the metallurgical arts.

Lightning in Colonial and Contemporary Andean Religion

Lightning has a natural and cosmological association with twins in pre-Columbian and post-Conquest cultures of the Neotropics. The natural association refers to the close visual and auditory association or simultaneous occurrence of thunder and lightning in nature. Binary and triadic classification is prevalent in Inca culture, and this is apparent with regard to lightning, in which the classification involves both the lightning bolt and the light and sound (thunder) emanating from it (Garcilaso de la Vega 1960 [1609]:50–51; Gade 1983:772; Zuidema 1982:151–153). With respect to dualities, lightning represents a transcendence of sound and light. Light is related to the apprehension of categories, while sound refers to animating essences or life force (Classen 1993:18). Such concepts are also directly relevant to Inca cosmogony and to Viracocha the creator, implying the existence of earlier forms of lightning veneration, which predate the Inca expansions. Among some Aymara populations Viracocha is symbolically associated with thunder and thus has direct religious relevance to lightning (Demarest 1981).

Climatic phenomena have come to be more commonly anthropomorphized in the Andes since the Conquest (Condori Mamani and Quispe Huamán 1996:45, 157 n. 9). The syncretism of lightning by contemporary Andeans involves fusing the ancient concept with Roman Catholic saints, St. James (Santiago) and to a lesser extent Santa Barbara (Gade 1983:777–781).[38] Lightning's association with Santiago, while peripheral to the Spaniards, made him a formidable power and source of veneration to indigenous Andeans. The apotheosis of Santiago Matamoros as a lordly, omnipotent epiphany has no equal in the post-Conquest era. Andeans usually addressed Santiago with respect as patron, apostle, *señor*, or *tayta* or by the single appellation "Santiago Illapa." However, the bellicose demeanor and putative power of Santiago over all kinds of weather-related phenomena commands more veneration than adoration among Andeans (Gade 1983:377). The role of Spanish clerics in elaborating the Santiago cult appears to have been subordinate to that of indigenous Andeans, whose folklore and legends in the colonial perception are said to often confuse agency and cause.

Guaman Poma (1980 [1583–1615]) mentions that town folk sequestered a mother and her newborn infant with a harelip because this deformity made him a "son of Santiago." An old man of the village guarded the infant and the mother was confined and prohibited from speaking, looking at the sun or moon, and eating salt, chile pepper, or meat for a month (Gade 1983:779). Local townsfolk then flogged her with ignited straw ropes as a form of exorcism. She fled with her offspring into the hills and remained there for several weeks, until the infant died. The dead baby was then tied on top of a live black llama, which was sacrificed (see **Figure** 15). Both the infant and the llama were then buried together and sacrificed to "Rayo Santiago." The mother returned in ceremonial clothes, and the occasion was marked by music, singing, and a five-day drinking binge primarily involving consumption of maize beer or *chicha* (Gade 1983:779). The previous description suggests that syncretism between indigenous beliefs surrounding Illapa and babies born with certain

---

[38] Like Santiago, Santa Barbara has played an important role in highland Andean folk religion from the sixteenth century onward. Unlike the dry-season date of July 25, which marks the fiesta of Santiago, her feast day, December 4, falls at the beginning of maximum thunderstorm activity in the central Andes (Gade 1983:781). Lightning was expected to be particularly fierce if the saint's feast day falls on a Tuesday or Friday, the two demonic days of the week in Andean folklore.

abnormalities and twins were absorbed early on by indigenous Andean populations.

The current perception is that Santiago "makes" thunder and lightning and sends them to earth. Santiago is also considered the protector against lethal lightning bolts, which strike with regularity in some regions of the Andes, particularly the high sierra (Gade 1983:778–779). Lightning is revered more on the high plateau than elsewhere. Places where lightning is known to have struck are marked with respect, for Andeans believe Santiago has come to earth there (Gade 1983:780). The cult to St. James (Santiago) was particularly apparent in the northern Bolivian altiplano. Since pre-Hispanic religion had among most cultures anthropomorphized lightning, the acceptance of a new epiphany, which had similar meaning and significance to spiritual veneration, involved no structural break or perceptual disjunction. By the early Colonial period, the indigenous mind had already begun to fuse Illapa into Santiago (Morúa 1922:230; Bertonio 1879 [1612]:278; Guaman Poma 1980 [1583–1615]). The timing of the feast day of Santiago, July 25, conformed to Spanish Catholic tradition. Guaman Poma (1980 [1583–1615]:34) felt Santiago had earned the right to be venerated in the religious calendar of the Andes by virtue of his role in "saving the Christians," thus his feast day was declared a Pascua—a religious festival extending over several days (Gade 1983:780). All-night vigils were commonly held, and offerings of burnt coca, libations of maize beer or *chicha*, llamas, fasting, and sexual abstinence were common during Colonial period festivities (Guaman Poma 1980 [1583–1615]).

Andeans invoke Santiago on his assigned day to intervene against hail, lightning, drought, and unseasonal freezes. Santiago is the patron saint of various high-altitude Aymara villages and in some cases is important enough to have a side altar in the local church. In some regions of the Andes, Santa Barbara held a closer association to lightning. The feast of Santa Barbara is December 4, at the beginning of the part of the year when thunderstorms are most intense in highland Bolivia. This was an obligatory day of rest in some Andean communities due to fear that Santa Barbara would send a bolt crashing down to punish those who dared to work in their fields (Costas Arguedas 1967/2:211).[39] Numerous chapels and parishes with her appellation attest to Santa Barbara's high position in Andean religion, although her popularity can also be attributed in some regions to her role as patron saint of miners (Gade 1983:781).

Lightning in pre-Conquest times generally takes on a triadic organization. Some sixteenth-century accounts describe this spirit as a trinity com-

[39] Such legends and folklore reinforce an association between Illapa and Viracocha the creator (see Pease 1978; Sullivan 1985).

posed of father (Chuki Illa) and two twin sons, Katu Illa and Inti Illapa (Polo de Ondegardo 1916 [1561]:6; Acosta 1962 [1589]:2210). Such triadic associations have come to take on an anthropomorphic emphasis in the Colonial and post-Colonial periods. For example, hail (*chikchi*) was perceived as anthropomorphic, three brothers, who were endowed with malevolent and terrible knowledge and power and intimately associated with crop failures and sterility (Condori Mamani and Quispe Huamán 1996:44–45, 156 n. 4). The association of sling stones with lightning may be related to their shape, which closely approximates hail. In both the modern and colonial cases, the spirit of lightning takes on an anthropomorphized triadic organization. St. James (Santiago) is often described as twin brother of St. John, or confused as the twin of Jesus Christ (Gade 1983:779; see also Krappe 1932:19).[40] These data suggest that climatic phenomena were perceived as manifestations of animating essences or life force that took on a dualistic and triadic classification and were symbolized by various forms. As indigenous Andeans combined the two identities into the single appellation Santiago Illapa—an association highly favored by analogy and metaphor—further cross-identification came from the convergence of dioscuric mystique into the Andean idea of magical twins. An important outcome of these cultural and religious accretions was to reinforce their attachment and validation of earlier indigenous beliefs. By simply invoking Santiago, Andeans were allowed to continue to hold lightning-related ceremonies long after the Conquest, including fasts and vigils that resembled those of the Inca period (Gade 1983:779).

Lightning in Inca Cosmology and Mythology

Lightning had important significance in Inca mythology and cosmology, particularly with reference to the celestial realm. The *ceque* system in Cuzco was astronomically aligned and marked solar and lunar passage, the cycles of Venus, and the lunar calendar of 328 days based on a twelve-sidereal-month cycle (Zuidema 1964, 1980, 1982, 1989, 1995, 2011; Magli 2005). The Inca tied their mythic origins to mythological heroes traced to a place called the Inn or House of the Dawn located near Pacaritambo and their political authority to

[40] The mythological dimension of St. James as the bearer of thunderbolts was established in Judeo-Christian orthodoxy long before the Spaniards set foot in the New World. The symbolic association of Santiago (St. James) with meteorological phenomena can be traced to the word *boanerges*, which means "sons of thunder," a biblical metaphor applied to the Apostles James and his twin brother John in an apparent allusion to their evangelical zeal (Cardinale 1983). The tenuous link of Santiago to thunder also came to extend to lightning, for where a thunderclap is heard a lightning flash inevitably precedes it (Gade 1983:777).

their mythological associations to the Island of the Sun in Lake Titicaca (Urton 1999:34–37; Staller 2008b). Pacaritambo refers to a village of the same name located twenty-six kilometers due south of Cuzco (see **Figure** 6), (Urton 1990:20, 1999). The Inca are said to have emerged from one of three caves or windows from a nearby mountain called Tampu T'oqo, or "house of openings or windows" (see also Polo de Ondegardo 1916 [1571]:53; Murúa 1922 [1590]). The original Inca consisted of four pairs of siblings or spouses. Juan Betazonas (1996 [1557]:14) states that the first ancestral spouses to emerge from the central window called "rich window" (*Capac toco*) were Ayar Cachi and Mama Huaco. On reaching the valley, the mythological ancestors went to the top of a sacred hill called Huanacauri. There Ayar Cachi, one of the original mythological Inca, took out his sling and hurled sling stones in four directions, and they struck so hard they knocked down the surrounding hills and created ravines (Betanzos 1996 [1557]:14). Thus the Valley of Cuzco was shaped and created by a sling and sling stones, primary symbols associated with Illapa in Inca religion. We may infer that sling stones were not solely perceived as weapons, but also symbolized the mythological creation and transformation of the Valley of Cuzco. The Inca linked their mythology to the heavens in that they perceived Illapa as a man made up of stars in the heavens with a sling in his right hand and dressed in shining garments, who gave off lightning when he whirled his sling or wanted rain (von Hagen and Morris 1998:125; Rostworowski and Morris 1999:792). The description has clear reference to the mythological Inca Ayar Cachi. The name Ayar Cachi means "ancestor salt" (Urton 1990:21, 1999). Garcilaso de la Vega (1963 [1609]: Book 1, chap. 18) explains that *cachi* means salt. Salt, the spice of seawater and that of Lake Titicaca, harbors symbolic connotations to seafoam, water, fluidity, and the circulation of underground or subterranean water, as is widely conceptualized among indigenous cultures in the Andes (see, e.g., Salomon and Urioste 1991:14–15; Urton 1981, 1990 1992). According to their mythology, the underground passages Manco Capac and Mama Quilla descended into from the Island of the Sun in Lake Titicaca to Pacariqtambo were part of a subterranean water system (Garcilaso de la Vega 1963 [1609]:30; see also Murúa 1922 [1590]; Guaman Poma de Ayala 1980 [1583–1615]). Subterranean water has clear metaphorical reference to the concept of Pachatira.

Lightning associations to later veneration of Illapa appear to have been influenced by the earlier religious symbolism and rituals related to the Gateway or Staff God, in the context of the spread of Tiwanaku civilization and the Wari state (see **Figure** 12b). The weather- and sky-related deity Thunupa was worshipped throughout the Titicaca Basin during the Middle Horizon period to bring coming rain, thunder, and lightning. Moreover, the so-called Guerrero relief below and surrounding the image of Thunupa on the Gateway of the Sun has been found by scholars to record various celestial cycles. The natural world exists in multiple temporal cycles, some tied to an annual subsistence round and others to the periodicity of cosmological and mythological (epochs or world) cycles (Zuidema 1982:159–161; Rostworowski 1986:31; Classen 1993:143, 194–195; Urton 1999:40–44; Sullivan 1996:27). Thus multiple modes of a historical and mythological past are "real" in that they coexist and continually contribute to the ongoing process of life (Zuidema 1982; Allen 1988; Sullivan 1985, 1987, 1988). Andean periodicities have particular reference to sacred places, *huacas* and *ushnus*, which in some cases were used to calculate celestial cycles, particularly those associated with the coming of climatic or seasonal events (Zuidema 1982, 1995, 2011). Lightning was the major theophany of weather in Inca religion because it was believed to control thunder and, by extension, all climatic forces, particularly rain, hail, and rainbows (Rostworowski and Morris 1999:792).

Lightning and the celestial realm also play a central role in Inca cosmology. In a colonial document dated to 1613, a native Andean Pachacuti Yamqui drew a diagram of Inca cosmology as represented in the Coricancha or Temple of the Sun in Cuzco. The upper portion of this diagram includes a wide array of celestial as well as meteorological phenomena. Some Inca *huacas* were astronomical and marked relevant astronomical phenomena with natural features along the horizon to calculate the solstices on the basis of lunar and solar cycles (Magli 2005:26; see, e.g., Zuidema 1964, 1997). Divisions of time are directly related to the divisions of space, the most obvious examples being the seasonal cycles (Urton 1981:195). The rainy and dry seasons are related not only to the movements of the sun, but also to the position of the Milky Way, while periods or phases of the moon with respect to the position of the sun (*inti*) determine the planting of the crops. Therefore the structural and organizational properties that are projected into the sky result in a systematic integration of celestial bodies and cycles with the calendar of activities, and hence their integration into the overall Inca cosmology. In his Inca cosmology, the Inca Pachacuti Yamqui (1950 [c.1613]:266) separates the halves of his diagram by contrasting structure (male) and fluidity (female). The man and woman in the middle of the diagram present the focal point emphasizing the centrality of humans to the cosmos and also serve to separate gender-related aspects of the natural and cultural world (**Figures** 26a–b). This integrating and separating center provides an overall horizontal triadic organization to a vertical duality or union of oppositions (*tinkuy*). Male aspects of the cosmology are listed on the left side and include the sun, summer, rainbows, lightning, morning star, and so on, while females, represented by the moon, clouds, winter, spring, evening star, World Tree, a mythic cat, and so forth, are on the right (Classen 1993:Fig. 2; see, e.g., Zuidema 1982:159). The underlying principle is to bring balance (*ayni*) and harmony from the middle, the symbols of culture. Significantly, the black *Choque-chinchay* cat is situated directly opposite the male lightning and the rainbow.[41] This cat is a mythological figure in Inca culture usually described as a "crying" feline associated with lightning as a harbinger for rain and particularly dark clouds that appear just before a storm. Pachacuti Yamqui presumably included the cat because it was believed to also have a role in producing hail, and also because it was venerated in the Coricancha during different times in the annual cycle (Classen 1993:21). The cat continues to have similar associations in contemporary Andean culture. Urton (1981) has observed that shamans often use the contours and the darkness of storm clouds to make predictions about the intensity of a storm (cf. Magli 2005:29).

Rainbows (*k'uychi*) were seen as a harbinger for hail and lightning, and those visible during thunderstorms were called "cat of the *apu*" (Ccoa). Contemporary communities make placatory offerings to the cat of the *apu* because it is believed to control lightning and strike down both crops and people (Benson 1998). Mishkin (1940:237) found that in the community of Kauri in southern Peru, the Ccoa is referred to as the "cat of the *apu*" and, when visible "crawling up the side" of a rainbow, is a harbinger for hail and lightning. Here the black cat has three manifestations—female gender, male lightning, and rainbows—and the fact that it transcends an oppositional duality suggests the transfer of divinity through the vertical axis (*axis mundi*) (Staller 2008b; see also Classen 1993; Roe 1995; Benson 1998). Lightning appears in three forms: *chuquilla* (thunder), *catuilla* (lightning), and *intiallapa* (thunderbolt). The Yamqui diagram thus provides symbolic reference to feline imagery, mountain spirits, and natural forces controlling climate and affecting agricultural productivity.

Rainbow is defined in a number of Quechua dictionaries as both "rainbow" and "the seven colors of the rainbow." The notion of seven colors has long been fixed in Western meteorological lore, where it is considered the ideal numerical expression of basic astrological, alchemical, and religious symbolism. It is uncertain, however, whether seven and its associations are inherent to Incaic or pre-Hispanic Quechua-speaking cultures (Urton 1981:87; see also Mariscotti de Görlitz 1973:208). Despite a distinct pattern of triadic significance, the importance of the number four has generally been emphasized with regard to ancient Andean religion (see Zuidema 1982, 1984, 1989, 2002, 2011). The Inca Empire had a quadripartite division, and Tawantinsuyu means "Land of the Four Corners" (see **Figure** 6). The Inca state divided its

[41] The black cat climbing up the rainbow is not only a harbinger for hail and lightning, but also evokes the underworld connections between lightning bolts that strike bodies of water and pregnant women who give birth to male twins. In Andean culture, pregnant women usually are advised to stay indoors during a lightning storm associated with rainbows, because it is believed lightning may cause miscarriages and twin births. Felines are generally associated with the underworld and thus provide a metaphorical connection through the tripartite cosmos (see Gade 1983; Urton 1981).

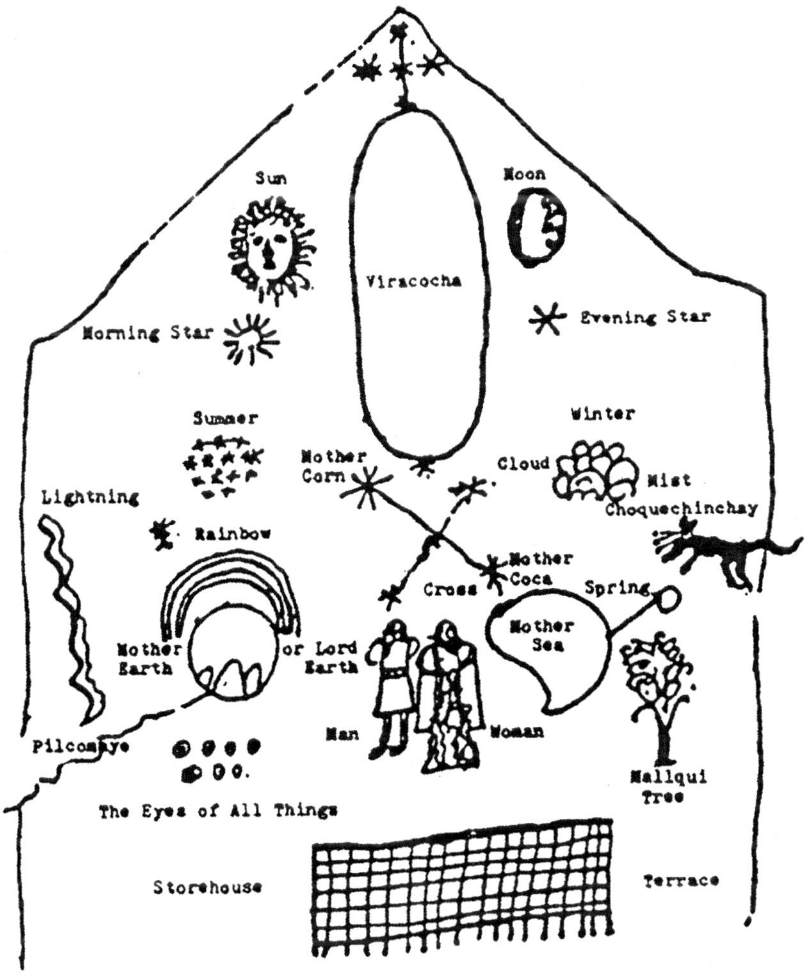

FIGURE 26A A diagram of Inca cosmology in the Coricancha by the Andean Jorge de Santa Cruz Pachacuti Yamqui c. 1613. Translation by Constance Classen. (From Classen 1993: Figure 2)

territories into four parts: *hanan* or "upper," associated with Chinchasuyu and Antisuyu, and *hurin* or "lower," associated with Collasuyu and Chinchasuyu (Magli 2005:23; Zuidema 2011).[42] Pachacuti Yamqui (1950 [c. 1613]:266), in his diagram of Inca cosmology, uses four lines to represent the

[42] The Inca divided into forty-two radial sectors the sight lines between the sectors, called *ceques*, on the same principle, and there were two *ushnus* associated with the main plaza Haucaypata: Hanan involved Inca rituals and rites in the annual calendar, and Hurin involved the Inca and their subject populations (Zuidema 1964, 1980, 1995, 1997, 2011; see also Bauer 1998).

FIGURE 26B Andean artisans have hammered the Pachacuti Yamqui cosmology into a metal (gold/copper alloy) replica of the drawing. The metallurgical arts have ancient religious and cultural association to lightning veneration. Note that the mythological black Choquechinchay cat's tail extends beyond the frame, surrounding the symbols of cosmology, as it does in the colonial Yamqui drawing, emphasizing its importance to legend and folklore surrounding Illapa. (Photo by Harold Hietala)

rainbow, as does the coastal Chimú culture at Chan Chan.[43] Sabine Hyland (2003:146) emphasizes that Blas Valera places much importance in the number

[43] The Chimú at the site of Chan Chan depict double-headed serpents emanating what appears to be lightning below four colored rainbows (see Figures 26a–b). *Spondylus* also figures prominently in their iconography, suggesting similar astronomical and religious significance, as identified in Inca culture. The primary deity venerated among the Chimú was the moon (Rowe 1964).

four by the Incas. However, the sun (Inti), thunder (Sallallaya), and lightning (Illapa) have three manifestations (Classen 1993:16). Similarly, there are male rainbows (*wankur k'uychi*), female rainbows (*k'uychu*), and androgynous rainbows (*urkuchinantin*); the latter are particularly dangerous to women (Urton 1981:89). Multicolored male rainbows usually appear high in the sky, their lower parts obscured by heavy clouds. Along the high, visible part of the rainbow, Andeans perceive a black cat (*ccoa*) that can be seen creeping steadily up the side of the bow. In contemporary culture, this "black cat" can be either male or female and is repugnant to Quechua speakers, who perceive it as a demon (Urton 1981:91–92).

Like the sun and lightning, the moon also has three names in contemporary Peru: Mama Quilla ("Mother Moon," or "Mother Month"), Mamacha (the Virgin Mary and [female] saint), and Coya/Colla Capac (Principal or Noble Queen) (Urton 1981:80). In the community of Kauri, the sun is related to agriculture through the channeling of fluids, and the moon (Colla Capac) is related to pastoralism (Mishkin 1940:235–236). Lightning is related to lunar cycles, rainbows, felines, lightning, and, more recently, springs and serpents. Quechua speakers describe rainbows as serpents (*amarus*) that rise out of springs (*pukius*) when it begins to rain (Urton 1981:88). Multicolored rainbows are visible during the day, while the black serpent is visible at night and associated with dark cloud constellations. Rainbow serpents have two heads whether they are thought of as circles or arcs. They rise from one spring and set into another. When they do this the two springs are said to be the "same spring." This may be seen as evidence that they are thought of as complete circles, which is how they appear high in the mountains (**Figure** 27; see color plate section). When rainbows move across the sky, they are thought of as willful, but most are malevolent (Urton 1981:88). They steal (usually from men) and, when associated with lightning, move into the vagina, causing great stomach pain. Nuñez del Prado (1970:88) states that when they rise from subterranean water they cause severe abdominal pain and headaches, vomiting, and sickness in general (cf. Urton 1981:89). In traditional culture, both men and women are forbidden from urinating when a rainbow is present because it can enter the stomach through the urine, causing severe stomach pains. When a rainbow enters the human body, a direct connection is believed to be established between the human body and the subterranean water within the *pukiu* from which the rainbow emerged (Urton 1981:89). Lira (1946:34) has stated that the waters from which rainbows arise have small colorful deposits (*colorcitas*), which can enter the body; if they do, one's urine will be rainbow-colored. All red rainbows can only be seen moments before and a few minutes after sunset (Minnaert 1954:182–183). The sun is thought of as male in Quechua sexual symbolism; red rainbows, however, may be classified as female because they "survive" beyond the setting of the sun (female = lunar = night). Rainbows can therefore be conceptualized as either female or male or a male-female combination. Drums are commonly played during the rainy season cultivation of maize, a period in the annual cycle associated with drums and male rainbows. Drums are also central to the symbolism of meteorological, seasonal, and agricultural phenomena in Inca times (Urton 1981:89–90).

The diagram by Pachacuti Yamqui thus provides symbolic reference to feline imagery, mountain spirits, and forces controlling climate and affecting agricultural productivity (see **Figure** 26a). The modern worldview perceives humans as occupying the interface, the terrestrial realm—what is called "this world" or "this life" by native Andeans. The cosmos is still maintained by creating a balance between the upper and lower worlds (Ossio 1973; Allen 1988; Maybury-Lewis 1989; Classen 1993). The duality expressed by this verticality is a basic tenet of native Andean cosmology and of great antiquity (Sallnow 1987:217). Dualism involving complementary oppositions permeates native culture from the way in which the landscape is conceptualized to the aesthetics of weaving (see, e.g., Valderama and Escalante 1977; Isbell 1978b; Cereceda 1987; Silverman-Proust 1988; Zuidema 1989; Classen 1990). Such dualism—as, for example, in the Hanan/Hurin (upper/lower) organization of the empire and the imperial capital, Cuzco—further extends to the institutionalization of natural resources and of spatially recognized social groups (Murra 1972 [1975], 1975b; Morris 1979, 1993; Zuidema 1982, 1983; Gelles 1995). Although not explicit in the ethnohistoric sources, Cuzco was divided into two opposing moieties: the upper (Hanan-Cuzco) and lower (Hurin-Cuzco); upper Cuzco (Hanan) was created by Viracocha and was symbolically associated with the male generating principle, mountain lords (*apu/wamani*), celestial deities, sun and lightning, royal palace estates, and the primary Inca *panacas* or lineages; lower Cuzco (Hurin) consisted of genealogically autochthonous non-Inca populations and their associated communities (*ayllus*) and was symbolically connected to Pachamama and thus to the female principle (Sullivan 1985:113–114). In the Inca cosmogony the lower Hurin societies were involved in the cultivating of the fertile bottomlands and the construction of terraces, and Hanan societies in the civilizing (*camachisqa*) of people (*camachicuna*) and lands (*patachana*) in order to create *ayni* (balance and reciprocity). Inca expansion and the spread of Quechua and Aymara throughout the Andes were intrinsically interrelated to the transformation of the landscape through human action, the creation of an ethnical order (Sullivan 1985; Covey 2006; Beresford-Jones and Heggarty 2011).

Earth, Sky, and Water in Andean Cosmology

Earth and water (Pachatira) is an important concept in Quechua cosmological thought. In some communities, Pachatira is considered the twin sister of

Pachamama (Casaverde Rojas 1970:150). Among some Andean cultures, Pachatira and Pachatierra are analogous. These descriptions refer to modern native Andean Quechua speakers, particularly in the Department of Cuzco, which therefore have some reference to earlier Inca culture (see Urton 1981:93; see also Nuñez del Prado 1974). Among indigenous populations Pachatira has come to be perceived as wicked in that she eats the hearts of men, who then die spitting blood. According to folklore and legend, Pachatira is generally associated with places in the natural world, hilltops, cliffs, and precipices; her preferred victims are adults and children who fall asleep in bad weather (Nuñez del Prado 1973:36; 1974:246–247). In this example there is an indirect relationship to lightning, death, and the night sky. Pachatira refers to a concept of earthly or subterranean fecundity in that all of the celestial animals are made up of "earth." So there is a relationship between animals and earthly or subterranean water, that is, the cosmic layers through their symbolic representations.

Symbolic associations of lightning to meteorological phenomena was expanded to the heavens and associated to mountain spirits or *apu*. One of the primary constellations associated with the heavens is the dark cloud constellation called Pachatierra, and the other is the celestial river or Milky Way. Pachatierra can sometimes refer to "dark cloud constellations," which are considered female and represent an ambivalent "transitional intermediate category of celestial phenomena" with reference to the upper realm (Urton 1981:175). The Milky Way is called "Mayu" and is of central importance to Inca cosmology, perceived as the celestial counterpart to the Vilcanota River (Magli 2005:26; Zuidema 2011). This symbolic association was so powerful that the water in the river was seen as the "same" water flowing in the celestial river and returning to earth in the rainy season (Urton 1981). The Inca are thought to have perceived the contours of dark cloud constellations, dark clouds of interstellar matter, within the Milky Way as symbolic of the contours of animals or birds (Urton 1981:88; Zuidema 2011). Rather than patterns forming celestial contours by connecting stars with lines, such dark areas usually had reference to different parts of the celestial river, the Milky Way, and thus were used in combinations with sight lines to refer to different parts of the annual cycle (Urton 1981; Zuidema 2011; Dearborn and Bauer 1995).[44] The dark cloud constellations are thus the celestial complement to the Milky Way and have come to have metaphorical reference to the underworld and agricultural fertility. Earth Shade or Pachatierra is a malevolent and perverse spirit that inhabits the lower world (Valderrama and Escalante 1977). Catherine Allen (1988:261) defines Pachatierra as a malevolent aspect of the Pachamama.

---

[44] The dark cloud constellations of the Fox and Llama are mentionied widely in the colonial accounts and appear to have been recognized on the northern coast by Chimú culture (Magli 2005:27).

Pachatierra therefore takes on a female aspect and is considered possessive and malignant. It resides in the earth in a localized place, such as cliffs and hilltops, rather than everywhere, as Pachamama does (Condori Mamani and Quispe Huamán 1996:39, 155 n. 6).

Pachatierra embodies an oppositional duality that is at once generative and destructive of many earth, mountain, and ancestral spirits. Pachatierra is not synonymous with earth or *tierra* because *pacha* denotes a temporal dimension lacking in the concept of *tierra* (Allen 1988:48; see also Staller 2008b:269–270). When lightning bolts cleave living rock or hilltops, the Inca would demarcate such places with rock carvings and cut stone architecture to denote their sacred qualities. Thus such practices have indirect reference to the ideas surrounding Pachatierra. The association of these constellations and such concepts to lightning veneration is through their associations with mountain spirits or *apu* as well as ancestral spirits (Nuñez del Prado 1974:247–248). Catequil, the lightning cult during the Colonial period, is associated with transformation, oracles, ancestor veneration, and various kinds of sacrificial and ritual offerings (Topic et al. 2002). Some rituals and rites also had reference to the Cult of the Dead as well as to recognition of the Inca state by their subject populations. The fact that some of these beliefs and perceptions surrounding Pachatierra and Pachatira continue to play a major role in contemporary indigenous culture throughout the Andes is an indication that such oppositional dualities with regard to their celestial associations as well as influence on perceptions of the natural landscape speak to Andean worldview and culture as a whole. In other words, they crossed linguistic and cultural boundaries in the Andes in prehistoric times and continue into the present among such societies.

Temporality of Lightning in the Andes

Lightning conveys the power and beneficence of all nature and was venerated to provide fertility to the coming agricultural cycle. In the Andes, weather and the forces of nature constitute a form of temporality closely associated with sacred rituals, dancing, and particularly music. The changing of the season is frequently seen to be the same alternation as the shift of primordial epochs reduced to a scale of annual duration. Consequently, songs that change the weather are often viewed as repeating the succession of primordial epochs by bringing about changes in the mode of being of the universe (Sullivan 1984:37, 1986). The origins of sacred music are associated with the regular rhythms of the seasons or the biorhythms of animal species, particularly felines and foxes, but the most dramatic expressions of sacred music and lightning are associated with life crises and rites of passage (Sullivan 1984:37–38). Each year around July, the entire Inca religious and political hierarchy would hold ritual celebrations in public, and private rites in Cuzco were focused on nature and its cooperation in the coming agricultural cycle (Molina 1959 [1575]:43–44; Zuidema 1989, 2008, 2011). Annual harvests would have been completed by this time of the year, so rituals, sacrifices, and offerings were focused on the next sowing (Gade 1983:775–776). Such veneration would hopefully conduce Illapa to send rain and keep away hail for the coming agricultural season.

Black, white, yellow, and red are sometimes associated with lightning veneration. In the Vilcanota valley of the Department of Cuzco, when lightning bolts strike and kill an animal, or maize stalks, offerings to Illapa included both red and white maize kernels. In this region, the Inca heartland, Andeans divide lightning into two color classes: yellow lightning, which is louder, stronger, injurious, and comes from the left side; and white lightning, which is perceived as good, brings good luck and lots of rain, and comes from the right side (Bolin 1998:50).[45] An indigenous curing rite for people who have survived a lightning bolt involves maize called *qhaqya sara* that is believed to have been struck by a bolt (Valencia Espinoza 1979:84–85). Such rare maize is dried and stored sometimes for years in the rafters of the hut, until someone is struck by lightning. When this happens, curers or *curanderos* rub the special maize cob over the body of the victim and command the malevolent force to leave the person. The medieval custom of apotropaics survives in the Andes, particularly the high sierra (Gade 1983:782).

Black has metaphorical references to the deep waters of the ocean (Ticcicocha)[46] and to the name given to two underground springs in the center of Cuzco (Gade 1983:775; Zuidema 2011). Zuidema (1992:23) states that the ocean was seen as surrounding and supporting the earth and that black llamas were related to the night, death, the ancestors, and the world outside Cuzco. Black llamas or those with black-and-white-spotted fur were often sacrificially offered to lightning as well as Inti, the sun (see **Figure** 15). Black and white maize kernels were used to make beer or *chicha* and commonly offered to the Illapa idol at certain festivals and rites (Staller 2006:454, 2010a). Cobo (1990 [1653]:166; vol. 1, chap. 35) mentions that in Inca purification rituals during the Citua Raymi festivals maize was used to purify living space and in healing rituals. The Inca would make a flour made first from black kernels and then white kernels. The walls and floors were then scrubbed with some of this maize flour while some was burned at the same time. Curing rituals also involve maize

[45] Black and white cords on *quipus* appear to indicate complex dualities such as subtraction/addition, increase/decrease, abundance/dearth, etc. (Platt 2002:247). This suggests these colors played a major role in the administration of the empire and various commodities stored in administrative centers along the Inca highway.

[46] Ticcicocha also has reference to the Inca belief that after sunset the sun followed its underground road through the deep waters of the ocean. Ticcicocha, "lake of origin," also referred to two natural springs in the center of Cuzco. These springs released their waters into the Hanan *ushnu* beside Haucaypata, the main plaza (Zuidema 1980, 1992, 2011).

flour made from white and black kernels that was mixed with crushed seashells of various colors. The flour mix would be put in a sick person's hand, and then, chanting certain words, the curandero would blow it as an offering to the *huaca*. The transfer of divinity was through distinctly colored varieties of maize within a certain prescribed sequence; sight (color), touch, and fluidity (smoke) mirroring a complementary duality in the cosmos (structure/fluidity) that maintains the boundaries of the senses (Classen 1993; Staller 2006). These data suggest that the conjunction of black and white possibly inferred divinity or access to different cosmic layers through a transference of divinity (see also Staller 2006:454).

Some (black-and-white) snakes symbolize lightning bolts and are linguistically related to the surrounding symbolism in that they mimic to a certain degree what still happens in certain rituals, in which black-and-white popcorn maize varieties are used in curing or made into maize beer during festivals and rituals dedicated to Illapa (Cobo 1990 [1653]; Staller 2006, 2008b, 2010a).

Serpents appear to also have had associations with lightning on the coast, as evidenced by Chimú iconography at centers such as Chan Chan (**Figures** 28a–b). These colors also have reference to the transfer of divinity in various Inca rituals (Zuidema 2011). Such veneration would hopefully conduce Illapa to send rain and keep away hail for the coming agricultural season. Thus lightning was venerated to provide fertility to the coming agricultural cycle (Gade 1983:775). Knowing when to plant and harvest, when the rains were coming, or when drought was beginning would have been invaluable to sustaining human societies in the face of climatic extremes. Individuals considered to have such esoteric knowledge were seen as distinct from other persons and imparted status or prestige, as was the case of lightning (Molina 1959 [1575]:29; Murúa 1922 [1590]:234). The repeated triadic pattern associated with the transfer of divinity, climatic phenomena, and their religious cults may be a reflection of the centrality of the tripartite cosmos to pre-Hispanic religion. Esoteric knowledge surrounding meteorological phenomena would have been of critical importance to the survival of Andean populations dealing with such an extreme, unpredictable, and in some cases inhospitable environment. The linguistic evidence indicates a clear religious and ritual association of lightning with twins and a triadic classification.

Rites of passage mark off the time line of an human individual life cycle, and certain rituals and lightning *huacas* focused on astronomy were central to the Catequil cult of lightning in the Late Horizon and early Colonial period. Each December, Illapa was venerated during the Inca festival of Capac Raymi. According to Acosta (1962 [1589]:268), three statues of the Illapa trinity were brought out of the Coricancha with the three idols that represented Inti or the Sun. Here again we find a triadic structure associated with the primary icons of Inca veneration. Penitence and blood sacrifice were central to the rituals and rites surrounding Capac Raymi (Cobo 1990 [1653]). Sacrifices included

FIGURE 28A Chan Chan, the Chimu capital: adobe friezes of snakes breathing lightning into one another below a four-banded rainbow. Such images decorated the adobe brick walls in labyrinth of corridors and elite compounds or *ciudadelas*. Chan Chan is largest-known city in the ancient Andes, with elite compounds covering an area of six square kilometers. Chan Chan covered an area of over twenty square kilometers, and many of the adobe corridors, tombs, temples, and compounds or *ciudadelas* were decorated.

FIGURE 28B Symbolic associations of rainbows and snakes in Chimu culture may also reflect earlier cultural concepts related to Pachatira and the spiritual forces of procreation and fecundity within the earth. The iconography suggests that such perceptions played an important role in pre-Columbian religion along the North Coast.

children—males born feet first (*chaqpa*), with cleft palates or harelips (*qhaqya sinqa*), or with head hair (*chaki wawa*), as well as single-sex twins (Illapa *kuri*), who were thought to have been divided by a *rayo* or thunderbolt (cf. Gade 1983:776; Murúa 1922 [1590]:234; Mariscotti de Görlitz 1978:366; Besom 2010:401; see also Cobo 1990 [1653]; Guaman Poma 1980 [1583–1615]). These sacrificial victims were later buried at a special place, often mountain summits or near or with *huacas* (Molina 1959 [1575]:94; Reinhard 1983, 1985; Ceruti 2004; Reinhard and Ceruti 2010). Sacrificed children during the Capac Hucha rituals and rites were usually interred on mountain summits (Duvoils 1976; Benson 2001; Staller 2008b; Besom 2010; Zuidema 2011; Reinhard and Ceruti 2010).

The Capac Hucha rite involved child sacrifice including twins and individuals with certain deformities as those outlined for lightning veneration. Capac Hucha was central to creating a spiritual and political relationship between the Inca state and their non-Inca subject populations. When being sacrificed, the most beautiful non-Inca children of the empire became *huacas*, and the places where they are buried—primarily mountain summits and lightning *huacas*—became synonymous with such places and their respective communities (McEwan and van de Guchte 1992; Besom 2010). Archaeological evidence along the North Coast in the form of effigy vessels suggest that individuals with such disfigurements were given status and venerated as early as the Early Intermediate period in association with Moche culture (**Figures** 29a–b). Such evidence provides additional supporting evidence of the possible existence of earlier ritual practices and rites of passage that were later associated with "sons of lightning" and, moreover, that such males were ascribed status within the society. The term *villca* (Aymara) can mean both "sun" and "shrine" (Bertonio 1956 [1612]:386). In the seventeenth century, the term meant someone who by achievement or marriage entered into the society of *huacas* —a very important individual who partakes of the status of a *huaca*, a superhuman person (Salomon and Urioste 1991 [c. 1598–1608]:chap. 2 sect. 10 n. 44). Andean religious cults often involved veneration of apical ancestors, their patrilineages and *ayllus*, as well as large clanlike groups identified with major founder *huacas*—in the case of lightning, founding ancestors (Staller 2008b:293–294). Thus such sacrificed victims were given special status as *huacas* (i.e., sacred and extraordinary), and special status was extended to their communities or *ayllus* (McEwan and van der Guchte 1992; Staller 2008b; Besom 2010). Capac Hucha sacrificial victims appear to have been initially given their rites of passage into *huacas* in veneration of the sun or Inti (Zuidema 1989, 2011; McEwan and van de Guchte 1992; Reinhard and Ceruti 2010).

FIGURE 29A Moche effigy vessel of Cut Lip, a high-status elite Moche male. This individual was depicted in three different portraits on effigy vessels differentiated by their headdresses. Serpents have symbolic associations to lightning, twins, and certain disfigurements in the Andes, suggesting that such cultural associations may be of considerable antiquity on the North Coast of Peru. (From Donnan 2004: Figure 8.8)

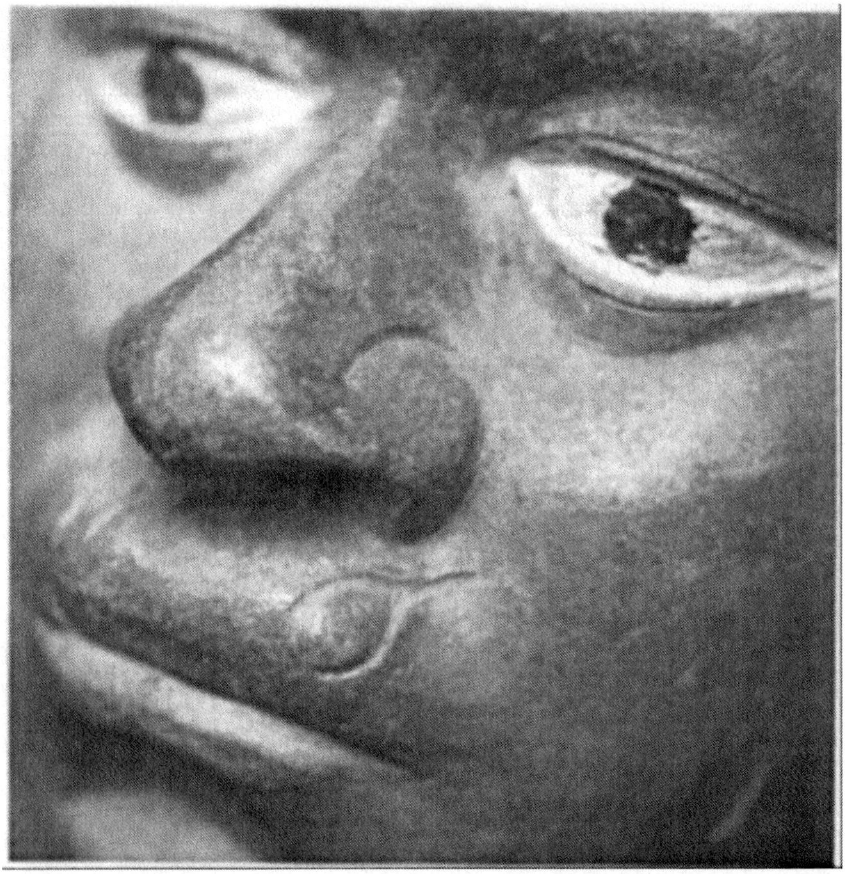

FIGURE 29B Close-up of cut or lip deformation. Moche effigy vessels of male twins and individuals with cleft palates have also been identified. Such effigy vessels have broader implications for the antiquity of such cultural concepts and the various symbolic associations to lightning and ascribed status to "sons of lightning." (From Donnan 2004: Figure 8.5)

It appears according to the account of Bernabe Cobo that Capac Hucha child sacrifices were also involved in certain rituals and rites with reference to Illapa and that lightning had increasing associations to earthquakes. A possible explanation for such changes may be related to the catastrophic earthquake in 1650, which leveled most of the colonial architecture in Cuzco, including the monastery of Santo Domingo. The Dominicans had built their monastery over the Coricancha or Golden Enclosure, which survived the quake unscathed. According to various chroniclers after this catastrophic event, Illapa became increasingly associated with earthquakes and certain Capac Hucha rites were expanded to include the primary trinity of the moon, sun, and lightning (Staller 2010a:16; Cobo 1990 [1653]; see also Acosta 1962

[1589]:268; Anonymous 1906 [1580–1621]; Polo de Ondegardo 1916 [1561]). Such associations may also have been inspired in part by what were seen as fictive kin relationships between various Inca emperors and lightning *huacas*, and between certain lightning *huacas* and Pachacamac. The oracle at this important religious pilgrimage center in the Lurin Valley was perceived to have the power to predict the intensity and timing of an upcoming quake (von Hagen and Morris 1998). The western coast of Andean South America is located along a subduction plate and thus is highly prone to earthquakes, some of them catastrophic. Earthquakes and their association to Illapa in the Colonial period provided indigenous populations a religious basis for continued veneration of Illapa and its traditional rituals and rites (Cobo 1990 [1653]:126–133).

Lightning and the World Inside

Caves have been found in association with some lightning *huacas*, however it is water that is the primary element connecting lightning to the underworld or *mundo interior* (i.e., springs, lakes, ponds, lagoons).[47] The Catequil cult and religious beliefs and rituals invoke the male generating principle, which involves the channeling of fluids, particularly water through the landscape and agricultural fields. Such concepts are related to Inca cosmogony associated with the civilizing of the natural landscape and its people through the imposition of an ethical order (Pease 1973; Sullvian 1985). Water is seen to flow through the earth into the underworld and into the sky in the form of the Milky Way, which in Misminay, Peru, is seen as a reflection of the local river (Urton 1981:38). The Quechua disposition of heavenly space reflects the fact that their destiny is shaped by the manipulation of water, as evidenced by customs such as ritual washings, ritual beverages, and the manipulation of irrigation technology. Human knowledge affords control over items in this water that originate in and that define the nature of heavenly space (Sullivan 1988:118). The human ability to manage life-giving water stems from the symbolic foundations of human knowledge, based on what is no doubt ancient sacred symbolism. The importance of lightning *huacas* or Illapa *ushnus* to channeling fluids and water ideology is evidenced by the underground channels or canals and drains found with many of the truncated platforms as well as various carved rocks, fountains, and basins, which also channel fluids across their carved surfaces at such sites (Staller 2008b; Topic et al. 2002). Burger (1992:141–

[47] The Quechua term for the underworld realm is *uchu pacha* (Allen 1988:63). Unlike Mesoamerica, in the Andes and particularly among the Inca, references to the underworld were always metaphorical or indirect (Zuidema 1992:23).

142) has pointed out that the regulation of water drainage for platforms would have been critical to the stability of such solid structures, since saturation with rainwater would have been a primary cause for their structural instability.

Submersion into the underworld from this world is generally through certain locations that contain water such as springs, whirlpools, and rivers (Sullivan 1988:122; Urton 1981:90–93). Such places are also associated in Andean mythologies as places of emergence into the present creation cycle (Sullivan 1985). Among the Shipibo in the Upper Amazon of eastern Peru, one enters the underworld by passing through deep pools of water, entering caves, or penetrating mountains (Roe 1982:117). Mountain springs are conceptualized as being the locations from which animals originate (tutelary spirits?) as well as caves, the principal places of emerging or entry into Pachamama, the Earth Mother (Urton 1981:93). As Lawrence E. Sullivan (1988) observed, "The underworld houses the nadir of the universe, the point at which human life, despite considerable resistance is drawn. It may be the *terminus a quo*, the outermost limit of human perception, even if it is not the *terminus ad quem* necessarily sought by human beings" (Sullivan 1988:123). Attraction to the underworld through lakes, ponds, natural springs, and caves is related to ancestor veneration, the emergence into the existing creation by mythological ancestors. In contemporary culture, primarily shamans and *curanderos* are believed to have access to the *mundo interior*, which presumably provides access to cosmic and spiritual realms. Terms given meteorological phenomena and transformation resulting from climatic events, such as the cat of the *apu*, Pachatierra, *Illapa huacacuna*, and *Illapa ushnu*, further indicate that in this world such places are considered sacred and often involve penetration of cosmic layers through esoteric knowledge of the spiritual domain found there, which may also imply some kind of access through a center or particular sacred location in geographic space (Staller 2008b; Eliade and Sullivan 1987).[48] Among religions in which the worship of nature plays a prominent feature, radically different levels of existence are felt to interpenetrate and coexist (Sullivan 1985, 1987; Frazer 1926; Eliade and Sullivan 1987). Such interpenetration and coexistence is manifested in pre-Hispanic Andean sculpture and art. Some native Andeans have come to believe the underworld is inhabited by little beings, both human and animal, and local protective spirits, as well as by a deity of Christian derivation, the devil (Condori Mamani and Quispe Huamán 1996:24, 148–149 n. 16; Valderama and Escalante 1977). Such beliefs continue into the present

[48] The "center of the world" or *axis mundi* can take on a range of symbolic forms in the Americas, including Cosmic or World Tree, Cosmic Mountain, ladder, vine or rope and house pillars (Eliade and Sullivan 1987; Staller 2000–2002, 2008b).

throughout southern Ecuador, where male dwarfs are often accused of impregnating women when their husbands are working in the fields.[49]

During the religious purges by colonial authorities and Catholic missionaries, religious and symbolic associations to the underworld were harshly condemned as the dwelling place of the devil (Nuñez del Prado 1979:238–239; see also Polo de Ondegardo 1916 [1561]; Molina 1959 [1557]; Anonymous 1906 [1580–1621]; Arriaga 1968 [1616]; Cobo 1990 [1653]). Remnants of such religious and cultural beliefs continue, however. The people in the village of Qotobamba also see the underworld as inhabited by tiny animals and people who live an existence that mirrors life on earth (Sullivan 1988:123). The underworld represents a reverse of the terrestrial, a kind of complement: they are bound by a hostile or fearful symmetry. Complete existence in one implies incomplete existence in the other (Sullivan 1988:123–124; Eliade and Sullivan 1987). It is precisely this relational quality that allowed Claude Lévi-Strauss to brilliantly analyze the "science of the concrete": that is, the repetitive formal relationships embedded in deep structure of myth are also embedded in the serial presentation and formal relationships in mythic narratives (Sullivan 1988:124; Lévi-Strauss 1966). The mirror image is marred by time and therefore testifies to the imperfection of space in time. In the Andean worldview, it is impossible to live in both worlds at the same time, since they are spatially and temporally exclusive, yet they exist together in the same experience only in an ambiguous and negative way (see, e.g., Bateson 1972:271–278).

Lightning Shaman: Human and Animal Familiars

Any study of lightning among contemporary indigenous and Latino populations in the Andes and Mesoamerica requires some mention of shamans or what Spanish clerics referred to as "sorcerers." As has already been discussed in the previous sections, a primary concern of religious clerics and the colonial governments was the "sorcery" associated with such individuals and their perceived power to access the spiritual realms (Eliade 1964:4–7; Joraleman and Sharon 1993:4). The term "shaman" came to the New World with the Spanish, who referred to most all religious specialists in these regions as *chamán* (Sharon 1978; Jorelman and Sharon 1993). The word "shaman" initially comes to us through the Russian and based on the Tungusic

[49] John Staller spent over twenty months carrying out fieldwork in coastal El Oro Province and the southern highlands around Cuenca, Ecuador, and found such tales commonplace, particularly when males were away from home for several months, working in the fields. Such small people (mainly males) were said to affect the weather and to appear seemingly out of nowhere, because they were from the *mundo interior*. Native societies in this part of the Andes try to avoid direct confrontation at all costs, and such folklore provides a means of avoiding conflict within families.

word *šaman*, in reference to certain magico-religious specialists. This term is derived from the Pali *samana*, which is believed to have its roots in the Sanskrit term *šramana* (Eliade 1964:495).[50] The Chinese later incorporate the term as *sha-men*, a transcription of the Pali term. Some Western scholars later rejected this etymology, asserting that the Tungusic term was part of the Turko-Mongolian language family by virtue of the phonetic correspondences, and concluded that the Tungusic *šaman* was the equivalent of the Turko-Mongolian *kam* (*qam*) (cf. Eliade 1964:495; see Laufer 1917). Later linguists found similar terms in the Tokharian *samane* "Buddhist monk" and the Sogdian term *šmn* = *šaman*, implying that the term could also have Indian roots (Eliade 1964:496). Mircea Eliade's (1964) study of shamanism in different parts of the world and among various cultures indicates that aspects of shamanic practice are universal, embodied in a "shamanic cosmology" which to varying degrees has similarities and associations among indigenous populations worldwide (see, e.g., Wilbert 1987:149–150; Sullivan 1988:390–400). The existence of such a cosmology is reflected in part by the extent to which indigenous populations have come to associate certain religious specialists with the term. For example, most native Andeans refer to religious specialists, healers, and those individuals who are caretakers of the soul and have the power to access the spiritual realms as "shamans" (see Reichel-Dolmatoff 1972; Wilbert 1987; Sullivan 1988; Sharon 1978, 2000; Joraleman and Sharon 1993).[51] Quechua and Aymara speakers refer to religious specialists priests and diviners as *hanpiq* and sorcerers as *paquas*, but some Andean communities nevertheless appear to distinguish such specialists from a shaman (Sullivan 1988; Valderrama and Escalante 1977). In colonial and postcolonial times shamans, *hanpiq*, and *paguas* replaced the Inca rulers in their perceived power to communicate with oracles and *huacas* on behalf of their communities, towns, and individuals (Sharon 1978, 2000; Platt 1997; Staller 2006; Valderrama and Escalante 1977). Significantly, Santiago (St. James) represents the patron saint of curers and shamans in many different regions of the Andes, which perhaps explains why lightning shamans are considered to be particularly powerful and why black magic is not generally widely practiced by Andean shamans (Platt 1997:199–200; Sharon and Joralemen 1993).[52] The perceived reward for veneration of

---

[50] "In the other languages of Central and North Asia the corresponding terms are: Yakut *ojuna* (*oyuna*), Mongolian *bügä*, *bögä* (buge, bu), and *udagan* (cf. also Buryat *udayan*, Yakut *udoyan*: 'shamaness'), Turko-Tatar *kam* (Altaic *kam*, *gam*, Mongolian *komi*, etc.)" (Eliade 1964:4).

[51] The association of the term "shaman" with traditional religious specialists became increasing popular after most of the indigenous populations and Latinos had been converted to Catholic orthodoxy in the late seventeenth and early eighteenth centuries (Joraleman and Sharon 1993:4–5).

[52] According to Christian orthodoxy, St. James, as patron of curers and shamans, does not tolerate black magic or scorcery (Platt 1997:199). Andean *curanderos* and shamans appear to have centered

Santiago (St. James) in the Bolivian altiplano is good health for humans, animals, and food crops, as well as economic prosperity (Platt 1997:198).

Shamanism exhibits particular emphases and/or characteristics in reference to South America in general and the Andes in particular that are distinct to varying degrees from what occurs in other parts of the world (see, e.g., Eliade 1964; Sullivan 1988). These specializations include "communication" to the spiritual realm, through *huacas*, by sensing or identifying "primordial heat"[53] in the natural world, a mastery over fire, caretaker of the soul, and ecstatic flight (Sullivan 1988:409–418). Control of fire as a technique for spiritual transformation is considered by scholars of comparative religions to have been by the first or primordial shamans of mythical times (Sullivan 1988:414–415; Eliade 1964:206, 257, 373). Other important associations include an ability to transform into a variety of forms, some meteorological, in the context of soul travel or ecstatic flight (Sullivan 1988:391–392, 409–418). Among contemporary shamans ecstatic flight is perceived as a result of spirit possession usually a condor or raptor, which imparts "true knowledge" or increased intelligence through "communication" with the spiritual realm (Sullivan 1988:417–418; Platt 1997:200–201). Shamans have close symbolic and metaphorical associations to raptors, reptiles, and felines, particularly jaguar (*Panthera onca* L.) and puma (*Puma concolor* L.) (Sullivan 1988:441; Saunders 1989; Benson 1998). In the Andes, such religious specialists are provided gifts and offerings as well as veneration in return for their advice and particularly powerful shamans have in some places developed a cult following (see, e.g., Sharon 1978; Platt 1997; Joraleman and Sharon 1993). It is perhaps significant that such individuals and the importance of shamans is centered in those areas of the Andes where metal ore deposits are located and where there is the most evidence of metalcrafting and the metallurgical arts in general. Ethnographic evidence from northern coastal Peru indicates that contemporary shamans or *curanderos* sometimes achieved a cult following and, significantly, that "highland lakes near Huancabamba...are famous pilgrimage sites for *curanderos*" (curers or shaman) in the northern Andes (Jorelaman and Sharon 1993:4). These lakes, ponds, and lagoons are in the southern highlands of Ecuador and northern highlands of Peru, and some are situated near Catequil de Huamachuco, the primary *huaca* and oracle associated with lightning (Illapa) in the Late Horizon and early Colonial periods (Topic 1992. 1998; Topic et al. 2002). Some of these sacred places were discussed in previous sections, but the

---

their presumed power around curing and predicting the coming rains, and there is evidence from early chroniclers that this was their primary role in pre-Columbian times.

[53] What is commonly referred to as primordial heat—that is, the spiritual realm in whatever manifestations it may take—is perceived among most native Andeans as reference to a life force (*enqa* or *sami*), a spiritual or animating essence perceived to permeate all things (Flores Ochoa 1977).

importance of lightning to contemporary shamanism in the Andes in general is evident from various scholarly sources.

Worship of the natural world underscores the fact that the sacred can appear in many forms. Mountain lords or *apu* generally call shamans to their vocation in the Andes. Thus it may be properly defined as a divine calling, a result of spiritual forces through divine inspiration. This vocation signals some manifestation or indication that the individual's soul possesses some extraordinary capacity or sensitivity to altered states and symbolic images in which the invisible are made manifest in dreams, visions, states of altered consciousness, or meditation, sounds, and particularly ecstatic flight (Eliade 1964:110–116; Sullivan 1988:40, 414–419). Shamans use ecstasy to gain true knowledge—in other words, to experience primordial heat, the unseen supernatural condition of existence. Shamans take the form of beings of a primordial age when they fly (Sullivan 1986:28; see also Sharon 1976; Joraleman and Sharon 1993). They may take the form of a bird, a beam of light, lightning, clouds, or a breeze. In other words, a symbolic complex that includes weather-related phenomena (thunder, clouds, and breeze), birds, animals, and stones embodies and evokes a strong celestial association. The human experience of passage from one symbolic realm of the imagination to another brings all realms of time and space—all modes of being—into relation with one another. The shaman masters this process of mystical transition, which essentially constitutes human life and maintains the universe (Sullivan 1988:415).

Paez shamans are given the calling by thunder and are believed to turn into thunder and to turn rival shamans into jaguars (Reichel-Dolmatoff 1972:54–55; Eliade 1964). Jaguars represented the co-essence or familiars par excellence for pre-Columbian rulers throughout the Neotropics, and such associations continue into the present with regard to shamans and *curanderos*. This may have to do with the growl of the jaguar, which is said to sound like thunder, and the fact that they generally hunt at dusk and dawn. Dawn and dusk are perceived in indigenous thought throughout Latin America as the "split between the worlds," as they are neither day nor night (Eliade 1964; Sullivan 1985, 1988; Sharon 1976; Joraleman and Sharon 1993). These associations are also related to shadow, as these are the times of the day when shadows are most apparent. One of the most important characteristics of jaguar behavior is the fact that they are able to swim and even known to fish, and bodies of water symbolize a conduit to the world below. Jaguars are one of the few large feline predators that swim, thus providing a logical association with the underworld. In Desana, Colombia, "power objects" are provided to the shaman by "thunder" that may appear as a roaring jaguar (Reichel-Dolmatoff 1975:78). The most powerful shamans in Peru are those who have been struck by lightning three times (Mishkin 1946:469). The origins of shamanism are related in part to being able to anticipate or predict future events and to interpret the past using "true knowledge." These abilities or responsibilities often include weather control (Sullivan 1988:391). Thus we may infer that the Inca ruler had such knowledge and abilities because only they were believed to make *huacas* "speak" and mountain summits are not only seen as *axis mundi* but also have a direct symbolic relationship to lightning veneration throughout the world (Staller 2006, 2007; 2008b; see also Frazer 1926; Lévi-Strauss 1966; Sharon 1976; von Hagen and Morris 1998).

In understanding the natural world as an expression of the intrinsic relationship between matter and spirit, the shaman becomes a specialist in the care of the soul, but the transformation of spirit becomes visible through consumption by fire (Wilbert 1987:183–184; Sullivan 1988:416). The consumption of mythic beings is often perceived as a cosmic conflagration of fire, the force that transforms the spiritual and symbolic world. Why fire? The ability to control fire is uniquely human. Mastery over fire is intrinsically linked to the ability to directly communicate with celestial realms of primordial heat, as well as to the ideology of consumption associated with ecstasy and ecstatic trance (Sullivan 1988:415). In many mythological accounts the world over, human life, as opposed to that of other animals, plants, and spiritual entities, is characterized by the possession of fire. For the shaman, the master of fire, the symbolism of fire centers the world. The *axis mundi,* the means by which the tripartite cosmos is spiritually connected, is essentially an expression of burning desire—the cosmic process of consumption contained and controlled by culture (Sullivan 1988:415; Eliade 1964:206; Lévi-Strauss 1966; Eliade and Sullivan 1987). Associations of shamanism to lightning through fire, obsidian, and ecstatic trance are most clearly apparent in indigenous culture and ritual performance, in central, northwestern, and western South America (Eliade 1964; Wilbert 1987; Sullivan 1988; Sharon 1976, 1978; Joraleman and Sharon 1993). Lightning has direct symbolic associations to fire and to the metallurgical arts, which involve mastery over and control of fire in the working of precious metals.

Amphibian fauna also have important cosmological and cultural significance in the Andes and in some cases an association to lightning through the weather, the annual cycle and for purposes of divination. The principal toad is *Bufo spinulosus* (*Chaunus spinulosis* Wiegmann). Toads are also referred to as Pachakuti (turning of the earth), Saqra (devil), Pachawawa (earth child), and Jacinto (hyacinth). Toads are referred to as "devils" because they foretell bad fortune when seen (Cobo 1964 [1653]:353) and are used by witches for malevolent purposes (Roca 1966:45; Jorelman and Sharon 1993). The names Pachatuki and Pachawawa refer to their common habit of burrowing into the ground to hibernate within, and later emerge from, the earth (Roca 1966:42). The earth is alive during the onset of the rainy season as they emerge in great numbers. After their initial emergence their behavior (mating, croaking, and so on) is observed for divinatory purposes, that is, to predict the coming rains (Roca 1966:58–59). Therefore toads are associated with weather and indirectly with lightning and divination through the foretelling of the agricultural cycle (Urton 1981:180).

Fictive Kinship in the Cultural and Natural World: *Guaoqui* and *Wayqe*

Ritual specialists, priests, diviners (*hanpiq*) or sorcerers (*paquas*), shamans (*chamán*), and curers (*curanderos*) communicate with the natural world through oracles or at *huacas* on behalf of their constituents, communities, towns, and villages (Valderrama and Escalante 1977). Powerful shamans were given gifts and offerings veneration and, if they were particularly powerful, developed a cult following (Sharon 1974; Allen 1988; Topic et al. 2002). Thus communication was through animating essences (*sami* or *enqa*) by Andean ritual specialists and their associated communities or *ayllus*. The concept of *ayllu* is neither straightforward nor simple regarding community or place; as Allen (1988:107, 257) points out, *ayllus* derive their well-being from the "same locality" or place and through this shared relationship are set apart as a distinct social unit (Allen 1986, 1988). Salomon (1991:21–22) defines *ayllu* of the seventeenth century as a "named landholding collectivity, self-defined in kinship terms, including lineages but not globally defined as unilineal, and frequently forming part of a multi-*ayllu* settlement." The concept is one of relatedness and not an entity of any specific dimensions or limits of scale (see also Urton 1990:22–23; Staller 2006:449, 453). *Ayllu* therefore denotes a relationship between people that can be expressed though a place or locality, descent, or political affiliation (Gelles 1996:8 Staller 2006:453). Modern community members or *comuneros* also emphasize an identification with and membership in a particular community (Condori Mamani and Quispe Huamán 1996:80, *f.* 17). Members of an *ayllu*, like *comuneros*, had claims on the *ayllu* resources—lands, camelid herds, and human energy in the form of labor services (*mit'a*). Agricultural tasks and the construction, maintenance, and feeding of *huacas* were carried out by *ayllus*, but provisions of food and maize beer (*chicha*) were expected in return (Murra 1975a [1972], 1980, 1982; Morris 1979, 1993; Morris and Thompson 1985). This type of reciprocity linked households, communities (*comuneros*), and their corporate kin groups to their leaders (*curaca*) and to the larger polity (Staller 2006:453). Reciprocity and redistribution formed traditional ties that bound culturally related societies into a social and economic unit (Rostworowski 1977, 1999; Murra 1980, 1985a 1985b; Zuidema 1982, 1983, 2011; Allen 1986). Communities and the larger polity were tied to *huacas* in some cases by "fictive" bonds. *Curacas*, ritual specialists, and nobles of larger polities were considered to have access to these spiritual and supernatural domains associated with a particular place in the landscape (Zuidema 1983:53, 1989; Urton 1985:253, 255–257, 1990:24).

Lightning played an important role in religious and cultural beliefs surrounding fictive kin relations between Inca rulers and lightning *huacas* and other sacred places. Inca rulers beginning with Pachacuti Yupanqui had fictive kin ties to Illapa and its *ushnus*, oracles, and *huacas*, adopting this force of nature as his brother or alter ego (Bolin 1998:46; Salomon 1987; Salomon and

Urioste 1991 [c. 1598–1608]; von Hagen and Morris 1998). Chroniclers state that the Inca ruler Tupac Yupanqui claimed four sanctuary-oracles as "sons" (Hemming and Ranney 1982:165). Colonial accounts also mention that the lightning *huaca* Cati Quillay could make other *huacas* "speak," including Llocllay Huancupa, one of the oracle "sons" of Pachacamac (Topic et al. 2002: 307; Rostworowski and Morris 1999: 796). Pachacamac was worshipped throughout the Andes because one of its powers involved predicting and controlling earthquakes (see **Figure** 12a) (von Hagen and Morris 1998:157). During the Conquest and later in the Colonial period, Illapa came to increasingly have symbolic associations with and perceived power over earthquakes. Such changes in lightning veneration were inspired in part by beliefs that some lightning *huacas* were *yañca*, that is, could make other *huacas* speak, and because such presumed power would evoke the oracle at Pachacamac, thus further reinforcing traditional rituals and rites to Illapa or Santiago.

The association of Santiago to lightning is related to the Spanish Catholic orthodoxy, where the emphasis had been on linking Santiago to thunder and twins. Such symbolic associations were immediately apparent among the native population, thus the combining of the names to "Santiago Illapa" among indigenous and Latino populations in the Andes. In the Andes, the notion of familiars is clearly apparent in the form of fictive kinship to the natural world, particularly meteorological and related phenomena, the cosmic realms, and sacred places (*huacas*). The Quechua terms *guaoqui* and *wayqe*, standing for brother or alter ego, were commonly used to speak of such familiars. Not only does the concept of familiars exist, but some individuals, particularly shamans and *curanderos*, have lightning as a familiar. Pachacuti Yupanqui, an important and powerful Inca ruler, elevated Illapa (lightning, thunderbolt) to the status of a major deity and adopted him as his *guaoqui* or *wawqi* (Bolin 1998:46). Such fictive kin relations were usually through the Sapa Inca, since the ruler was seen as divine, a living manifestation of structure, of light energy represented by the sun or Inti (Staller 2008b).

Pachatira: Rainbows, Serpents, and Water

Animal familiars are also associated with the annual cycle. For example, the December solstice is associated with celestial and terrestrial foxes. The exploitation patterns of the fox are governed by lunar cycles. The word *atoq* (fox) has a free association with *wañumarka*, which means "storehouse of the dead" and is the name of the principal *apu* of Misminay, Peru (Urton 1981:70). The *apu wañumarka* is full of foxes during the winter solstice because foxes give birth four days after Christmas, which is when they celebrate the winter solstice (Urton 1981:70). The association of the moon and women (Mama Quilla), is of course neither unique to the Andes, nor a totally arbitrary association in that it also involves the stars (i.e., lunar zodiacs; see Urton 1981:79, 1985; Zuidema 2011). Symbolic associations or referents among familiars and meteorological phenomena are apparent when we consider that rainbows are usually thought of as giant serpents that rise out of subterranean springs, pass through the sky, and then bury one of their two heads in a distant spring. Rainbows are therefore a manifestation, in reptilian form, of the forces of procreation and fecundity (*pachatira*) within the earth. Thus all meteorological phenomena are reflections of the presence and action of powerful natural forces, primarily forces associated with water, the source of fertility (Urton 1981:93; Sullivan 1985).

Animal familiars are also linked symbolically to celestial phenomena (Zuidema 2011). Dark constellations in the center of the Milky Way or celestial river are, during certain points in the rainy season (October through April, especially December through February), referred to as "the serpent" when visible at night during the rainy season. There is an association among dark constellations, serpents, and water. Since meteorological serpents (rainbows or *amarus*) appear only during the rainy part of the year, they exhibit a seasonal activity cycle similar to that of terrestrial reptiles. During the dry season the dark constellations are perceived as existing below the ground. Water in the form of rain returns to earth, where in its continuous cosmic cycle it flows in the tributaries of the terrestrial streams. In this way, the celestial and terrestrial rivers act in concert to continuously recycle water, the source of fertilization, throughout the universe (Urton 1981:172). The prevailing cosmology in the Andean world sees the earth as a flat or slightly curved mirror that alternately reflects the celestial and underworld order across and through itself (Urton 1981:37, 63; Lechtman 1999:227; Staller 2008b:287; see, e.g., Ortiz et al. 2008: Fig. 6.8).

Andeans still communicate with mountain lords (*wamani* or *apu*) and earth spirits and also carry out healing rituals on behalf of individuals, families, and entire villages and towns (Condori Mamani and Quispe Huamán 1996:35, 153 n. 5; Sharon 1972, 1976, 1978; Jorelman and Sharon 1993). Some *paquas* fulfill various roles, including diviner, healer (*hanpiq*), and witch (*layqa*). They are particularly well versed in and knowledgeable about the weather and astronomy as these phenomena relate to planting and harvesting of crops (Sullivan 1988:390–391). Such *paquas* are referred to as fieldmasters (*chakrakmayuq*), literally "he who orders in cultivated fields." These ritual specialists read the stars or coca leaves to determine the particular day that a community should begin planting or harvesting (Valderrama and Escalante 1988:230–233; Sharon 1978). In pre-Hispanic times such activities were generally carried out in sacred centers or in the ceremonial precincts of cultural centers (see Staller 2008b). Whereas shamanic practice and ritual in northwestern South America encompasses divination, curing, and black magic, such Andean practitioners are generally involved in providing agricultural fertility and predicting future events particularly as they relate to the sustenance and survival of associated communities (see, e.g., Sharon 1978; Roe 1982; Wilbert 1987; Sharon 1978; Jorelman and Sharon 1993).

Andean Lightning Stones

In some regions of the Andes ground stone axes are "thunderstones," and translucent stones, particularly crystal and obsidian, have close symbolic links to weather-related essences. Lightning and transparent stone crystals link the shamans to ecstatic journeys into celestial realm, and these are particularly powerful because thunder and lightning are often symbolic of death.[54] Shamanic separation of the body and soul is culturally perceived as death. There are thus metaphors for the powerful state of ecstasy that typifies the shaman's uniqueness and makes ecstatic flight possible (Sullivan 1988:409; see also Sharon 1978, 2000). Warao shamans control fire with their gourd rattles, which contain rock crystals. Their ecstatic flight through altered states of consciousness is "awakened" by the primordial fire and celestial light in their rattles, thus the voyage of their souls in the context of ecstatic flight is perceived as spiritual awakening (Sullivan 1988:416). Shamans travel in this awakened realm of primordial heat along the shaft of their rattle, which is also perceived as a central axis or *axis mundi*. Ashánlinka or Campa shamans in eastern Peru perceive the ecstatic flight of the shaman's soul as visible by flashes of lightning: *ipóreki*, "he flashes," so all can see (Sullivan 1988:417). Thus ecstatic flight among some cultures becomes visible through lightning.

Symbolic associations of lightning to death through warfare and conflict are particularly apparent during the Late Horizon. During the period of Inca expansion, lightning *huacas* and Illapa *ushnus* were often provided piles of sling stones as ritual offerings (Staller 2007; see, e.g., Lippi and Gudino 2010a, 2010b). Preliminary research at the equatorial lightning *huaca* of San Catequilla de Pichincha has indicated that sling stones may have been mixed into the clay used to construct a large circular platform on the summit (Staller 2011). The hilltop Inca *huaca* of Palmitopamba located in the *yumbos* on the western-side

---

[54] Rock crystals have strong associations to ecstatic flight, to bestow the power to rise to the sky (Eliade 1964:138–139; Sullivan 1988:416–420). There appears to be an empirical association with death by fire, particularly to *rayos* or thunderbolts that strike the earth or natural environment. Such associations are worldwide in relation to shamanic religions (Eliade 1964; Sullivan 1988).

cordillera have sling stone caches on the surface of the stone-faced terraces (Lippi and Gudino 2010a: Fig. 10.5a; 2010b: 627–628, Figs. 2–3). The use of sling stones in the construction of the San Catequilla lightning *huaca* and its importance as ritual offerings that were commonly placed on artificial terraces associated with truncated platforms such as Palmitopamba provide compelling evidence for their significance beyond their use as weapons. The various lines of evidence suggest that sling stones had mythological associations to Inca culture that involved the transformation of the Valley of Cuzco. Some chroniclers mention that sling stones were associated symbolically with lightning (Illapa) through death and Inca mythology. Such associations are particularly emphasized with the religious cult of Santiago during the Conquest and Colonial periods.

Apu or mountain spirits can take on many forms, both zoomorphic and anthropomorphic, or embody aspects of different things, which often have reference to the tripartite cosmos: for instance, the smiling deity or Lanzón at Chavín de Huantar and the Staff God at Tiwanaku. It is in such mountains that soul journey takes place during life cycle rituals, shamanic voyages, dreams, or altered states of consciousness because one gains access to the whole universe from the center (Whitten 1976; Sharon 1978, 2001). Mountain lords, *wamani* or *apu*, call Andean shamans to their vocation (Sharon 2001). The symbolic association of lightning to *apu* in the Catequil cult provides an indirect reference to lightning (see **Figure** 13). Transparent stone crystals appear during the Late Formative (c. 1200 BCE) period in the context of long-distance interaction involving, maize, ceramics, metal artifacts, and *Spondylus*/*Strombus* shell and expand to various regions of the Andes in later periods (Paulsen 1974; Villalba 1988; Lechtman 1976, 1984; Bruhns 1987, 1994; Temme 2000; Staller 2000–2002, 2007). Stone crystal continues to play an important role in shamanic rituals involving lightning and other weather-related phenomena. Polished or ground stone axes are perceived as thunderstones and are associated with World Tree mythology in the Upper Amazon (Roe 1982:140) and some regions of northern and eastern South America (Staller 2001). Many lightning *huacas* associated with the Catequil cult are crafting centers associated with the metallurgical arts with artificial terraces on the summits or the bases, infant burials in funerary vessels, and tightly flexed and upright subadult burials in association (San Pedro 1990 [1560]; Salomon 1987; Salazar 1998). Many have sling stones or stone cairn offerings and finely crafted marine shell as funerary objects in elite burials (Staller 2007). Maize beer (*chicha*), pottery, *Spondylus*/*Strombus* shell, and metallurgy are closely linked to Inca lightning veneration, and such associations continued into the early Colonial period. Another association involves the control of fire and the use of wind currents to sustain firing temperatures. In some regions of the Old World, metalsmiths are closely linked to shamanism and sorcery for the same reason (Eliade 1964).

Lightning in the Andes: Summary

One striking pattern to come out of this study of lightning is the degree to which all religious veneration has reference to the landscape, the natural world, or the celestial realm, and lightning is no exception. Places in the landscape struck by lightning were demarcated and venerated as *huacas* or sacred and extraordinary places. The four primary symbols of religious veneration during the Late Horizon period—the sun (Inti), the moon (Mama Quilla), lightning (Illapa), and Paqariqtambo—are all aspects of the natural world, be they actual, metaphorical, or mythological, as in the case of the latter. They were directly venerated as opposed to being or representing an aspect of a pantheon of deities or places of mythological origin. Lightning is still believed to cause twin births among various indigenous populations. The female *rayos* or bolts that strike the earth and in the case of pregnant women cause same-sex male twin births, or certain physical deformations such as cleft lips or harelips, became "sons of lightning" or *Illapa huacacunas*. These individuals were given special status and rank and protective rights. Linguistic evidence indicates that such effects and their results represented what would be perceived in Andean culture as a transcendence of a duality, and in Western eyes as an individual conceived without original sin—an immaculate conception. The cross or *crucero* had an association to mountain lords or *apu* and therefore to lightning and directly invoked the symbols and significance to the cult of Santiago. The subsequent importance of the Cross of St. James to the Order of Santiago clearly gave the Roman Catholic Church a means by which they could transfer veneration from Illapa to Christian saints and religious orthodoxy. The legacy of such conversion is nowhere more apparent than the great Catedral Metropolitana de la Inmaculada Concepción, the spiritual and culture center of Cuenca, Ecuador, where Augustinians and other religious clerics and the colonial governments suppressed veneration of Illapa and replaced it with veneration of Santiago. It is significant that lightning veneration involves the most complex folklore and legend in the central highlands of southern Ecuador and northern Peru and the Bolivian altiplano, as these were the regions with the most gold and silver ore deposits (Lechtman 1976, 1984, 1988; Hosler et al. 1990). The clear association of Illapa to the metallurgical arts is reflected in the archaeological record and in ethnographic and ethnohistoric accounts (Gade 1983; Bouysse-Cassagne 1993; Cruz 2009; Platt 1997; von Hagen and Morris 1998).

Worship of nature is also reflected in the fictive kinship relationships and symbolic association to and among places in the natural world; this was a central aspect of pre-Columbian religious worship in the Andes and other regions of the Americas. The ethnographic evidence indicates that veneration of the natural world continues in modified forms into the present. During the Late Horizon the Inca referred to themselves as children of the sun, and the ruler or Sapa Inca was considered to be the Son of the Sun, a living manifestation of eternal light (Staller 2006). The highest-ranking females, the *acllyas* and *mamaconas*, were symbolically linked to the moon (Mama Quilla) and appear to have been associated with some lightning *ushnus*. Recognition of Inca rule and political authority required support populations to honor such fictive kin relations and their symbolic correlates in the landscape, mountain lords or *apus, huacas, ushnus*, and so on—to make ritual offerings and bring tribute to sacred places that had symbolic reference to such animating essences (Rostworowski and Morris 1999; Staller 2008b).

Lightning is distinguished by the fact that the Inca state used such religious veneration to symbolically connect themselves to non-Inca subjects and their communities and polities through fictive kin relations. By making their subjects "Inca by privilege" in the context of child sacrifices or Capac Hucha rites, such non-Inca children *huacas* and their *ayllus* achieved status with the Inca state (Duvoils 1976; Besom 2010; McEwin and Van de Guchte 1992; Rostworowski and Morris 1999; Reinhard and Ceruti 2010). Such associations continue into the modern period; the Uru, an ethnic group in highland Bolivia, are referred to as *haqe huaca* by their Aymara-speaking neighbors because they are believed to have survived from primordial or mythological times (Manelis de Klein 1973:143). The principal lightning *huaca* at Huamachuco is also *yañca*—it could make other *huacas* "speak." According to chroniclers, the oracle at Pachacamac "told" the Inca ruler Tupac Yupanqui that the sanctuary oracle of Apurimac was one of its four sanctuary-oracle "sons" (Hemming and Ranney 1982:165). Significantly, chroniclers also mention that Cati Quillay, the lightning oracle at Cabana, was an emissary of the Inca and said to make other *huacas* "speak" (Topic et al. 2002:307). Cati Quillay is also credited with making one of the oracle "sons" of Pachacamac, the *huaca* at Llocllay Huancupa, "speak" (Rostworowski and Morris 1999:796; Topic et al. 2002:307). Therefore certain lightning *ushnus* or *huacas* were related to one another through fictive kinship, and communities (*llactas*) had fictive ties not only to their surrounding landscape, but also to the Inca state through *huacacuna* and Capac Hucha. Thus sacred places, like certain lightning *ushnus* or *huacas*, were perceived as having fictive kin relationships to one another (Staller 2008b:294; Besom 2010).

Jaguars are particularly important in lowland tropical regions, while pumas often have such symbolic significance in southern Peru and the Bolivian altiplano associated with lightning-related religious beliefs surrounding mountain lords and rainbows through folklore surrounding the cat of the *apu* and its perceived effects on twin births, and certain birth defects, which make such individuals sacred and extraordinary (Krappe 1932; Duvoils 1976). These characteristics symbolically link such individuals and their associated *llactas* or communities to those sacred places and mountain summits (see, e.g., Salomon and Urioste 1991 [c. 1598–1608]).

The Inca state perpetuated religious veneration to the sun, the moon, lightning, and the dead (mythological ancestors, dynastic rulers) as a means of legitimizing and sanctifying divine kingship and their political authority (Rowe 1946; von Hagen and Morris 1998). The Inca cosmogony also perpetuated an idea of an ethnical order imposed on the landscape and its associated cultures, which was readily comprehensible to its subject populations. Inca cosmogony indicates that lightning may have symbolic reference to intermittent cycles of chaos and destruction that separate the four creations or world cycles: the original mythological Inca were in some cases turned into stone, thereby symbolically connecting their culture to what was generally perceived in Andean culture as the oldest living matter on earth. Lightning also had clear associations to stone and to the creation of stone, as rock outcrops cleft by lightning were enclosed to mark their sanctity. Lightning bolts also had direct relationships to places of emergence, caves, natural springs, lagoons, lakes, and so on, that symbolically link the celestial to the underworld (Sullivan 1985). Illapa has a direct association to earthquakes through the principal lightning *huaca*, Catequil de Huamachuco, which was perceived as a *yañca*, as well as to fire by bolts, which cause forest fires or burn crops. These symbolic links provide a basis for understanding earlier associations to the creator, Viracocha, and to intermittent period of destruction and chaos, thus providing an ideological basis for earlier forms of such veneration.

Illapa was a polymorphic force of nature with symbolic and metaphorical references to Viracocha, creation cycles, Inca mythological origins, and intermediate periods of destruction or chaos. There is ethnohistoric and ethnographic evidence to suggest that Viracocha has associations to thunder among various Ayamara-speaking populations (Demarest 1981). Such symbolic relationships may be related to beliefs regarding creation cycles and Andean cosmogonies in general. A relationship between lightning and Viracocha is also suggested in Inca cosmogony (see Sullivan 1985). The ability of the Inca state to conquer and perpetuate its mythological and ideological authority over such a large geographic region over a relatively short period of time speaks to the linguistic homogeneity of this culture area and the extent to which such religious veneration and ideas regarding the spiritual realms were readily incorporated. Many natives continue to traverse an Andean landscape under the watchful eye of the Father Sun (Tayta Inti) and Mother Moon (Mama Quilla) and walk the high mountain passes (*apachitas*), beside lakes (*quchas*), across the plains or pampas, down warm valleys (*qhiswa*), into the jungle valleys (*yunka*) of the eastern cordillera in a mythic geography imbued with cultural and spiritual meaning (Gelles 1996:10; Staller 2008b:449–450). Lightning (Illapa) provided another means by which the landscape was sanctified and a basis on which communities gave meaning to the natural world around them. The connective patterns and traditions between Andeans and their landscape, and the role lightning veneration plays in such religious perceptions, clearly distinguish this region from what has been documented in Mesoamerica.

3

Mesoamerica

Introduction to Mesoamerica

Mesoamerica comprises the lower two-thirds of Mexico, all of Guatemala, all of Belize, and the western edge of Honduras (**Figure** 30a). It has long been viewed as a distinctive culture area located between Central America to the southeast and North America to the northwest. It is an area associated with diverse specific traits according to Kirchhoff (1968), and with multiple adjacent "high civilizations" according to Armillas (1957, 1964), including the civilizations of the Aztecs, Mayas, Olmecs, Zapotecs, Mixtecs, and Totonacs. Indeed, a number of cultural and linguistic characteristics help to define the culture area, such as the processing of maize with slaked lime (calcium hydroxide), maguey cultivation, writing, ball courts with rings, a 260-day ritual calendar, appreciation and use of jade, counting by twenties, a three-stone hearth, and several others (Campbell, Kaufman, and Smith-Stark 1986; Brown 2011). The number of ideas shared throughout large portions of this otherwise deeply diverse culture area is remarkable—all the more so in light of the fact that more than seventy different and mutually unintelligible languages are currently spoken in the region.[1] Among the concepts and related narratives shared throughout Mesoamerica are many connected to lightning and its ideological associates.

Mesoamerica includes large highland areas with expansive valleys and high volcanic peaks; Mexico's Citlaltepetl (Orizaba) reaches up to almost

[1] Most of the languages in Mesoamerica fall within the Mayan, Oto-Manguean, or Mixe-Zoquean language families, and the latter has recently been proposed as a separate branch of the Mayan language family (David Mora Marin personal communication 2011). An excellent overview of Mesoamerican languages can be found in chapter 11 of Carmack et al. 2007.

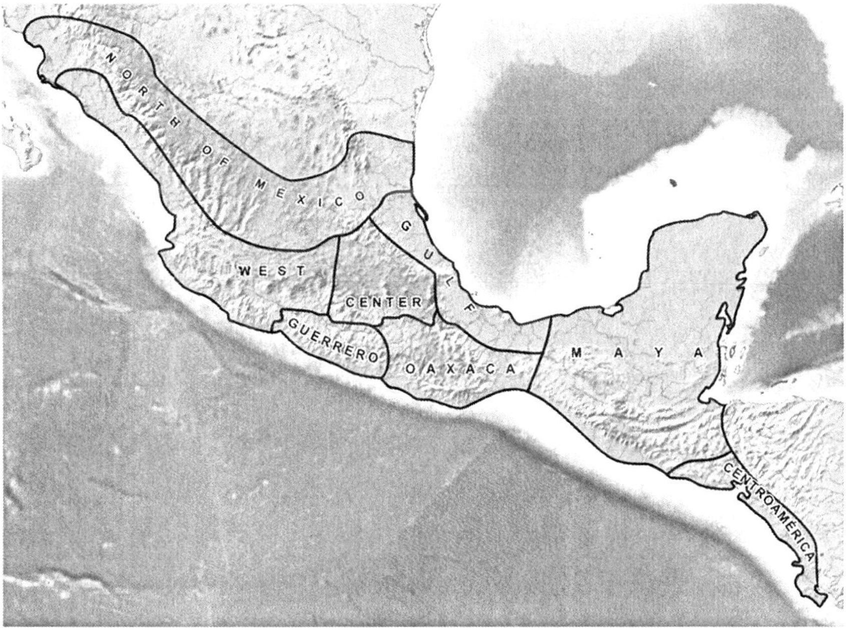

FIGURE 30A Mesoamerica shown divided into geographic subregions. The northernmost and southernmost named subregions are buffers, not part of Mesoamerica proper. (Courtesy of Wikimedia Commons)

19,000 feet, where lightning can frequently be seen lighting up the clouds. Tropical and subtropical lowlands below 3,000 feet make up about half of Mesoamerica's topography. Rainfall ranges from about ten inches a year in some locales to over one hundred in others, the greatest amount being found on both coasts flanking either end of the Isthmus of Tehuantepec, especially including the Mexican states of Veracruz and Tabasco, the Peten region of Guatemala, and Belize. A rainy season from May to October is complemented by a dry season between November and April. Mesoamerica's geographic diversity has created an environment in which places are separated from one another physically and ecologically. The ecological differences have united contiguous areas through the advantages of trade, resulting in considerable communication and allowing for the diffusion of ideas as well as trade goods.

Rainfall in Mesoamerica, variable in quantity and often unpredictable, has long been crucial to many aspects of life. This is particularly so, in addition to the vital role it has played in maintaining viability in the many ecosystems in the region, because of the dependence of maize agriculture on rain and the timing of its onset for sufficient yields to maintain human life. It is quite understandable, therefore, that rain deities, rain festivals, and rain dances were to be found throughout Mesoamerica, and it is likely that the rain and lightning deities of long ago were viewed as capricious then, just as they are today, by indigenous inhabitants of the region.

In Mesoamerica thunder, lightning, and rain are strongly associated with one another and with a deity who is sometimes called the lightning deity by scholars and some indigenous communities of the region. In other communities he is called the rain god, or the thunder god. Typically this deity lives in a cave on the mountain, and in various languages he is literally named the "Earth Lord," "Thunderbolt," "Mountain Lord," and similar epithets. The phenomenon of lightning, along with the deities, spirits, and co-essences (animal familiars,[2] *tonos*, or *naguals*) that are associated with him by members of the different linguistic groups, are variably perceived over time and from place to place. Nonetheless, the cultural expressions of these perceptions involve similarities supporting the conclusion that there is considerable cultural temporal and spatial continuity with respect to a number of features characterizing otherwise quite distinct worldviews. This in turn supports the proposition that there is a Mesoamerican culture area: an area sharing enough ideology, behaviors, and material culture to set the region apart from its neighbors in a meaningful way. The beliefs and behaviors surrounding lightning contribute substantially to this idea.

Many of the differences in worldviews among human societies can be attributed to three things: first, the varying ecological niches that different human communities inhabit; second, the unique histories through which these communities and the individuals belonging to them have proceeded; and third, the classification systems that different languages entail as guides for thought. With respect to the latter, a somewhat simplified picture of the different languages of Mesoamerica and how they are related will be helpful for placing in context the information to be presented below. The language picture in Mesoamerica today reflects the languages and migrations of peoples of the past, including three large and important language families (groupings of related languages) that could be related to three regions in particular at the time of the Spanish Invasion.[3]

---

[2] In Europe, particularly associated with witchcraft, the notion of an animal familiar involves a spirit or soul that is often embodied in an animal and believed to attend, serve, or guard an individual. In Mesoamerica the notion is further embellished by a belief in shared birth times and shared life fates due to a shared soul between the individual person and animal, and/or a belief that some persons, especially sorcerers, can transform themselves into their individual familiars. Furthermore, the familiar in Mesoamerica may be a meteorological phenomenon such as lightning that Westerners, but not Mesoamericans, would judge to be inanimate. When the term "animal familiar" is used here, it is in the Mesoamerican sense.

[3] Many individuals feel that "Spanish Conquest" is a misnomer and that the proper designation would be the "Spanish Invasion." It is hard to fault this line of thought, as the latter is both preferable and more accurate.

FIGURE 30B Mesoamerica map showing selected languages, cities, and archaeological sites.

The first region, focused on the Mexican state of Oaxaca, includes the languages of the Oto-Manguean language family, most of them tone languages, such as the Zapotec languages, Chatino, Mixtec, Chinantec, Mazatec, Ixcatec, Triqui, and Cuicatec (**Figure** 30b). Another branch of the Oto-Manguean family, the one including Otomi, Mazahua, and Pame, is located mostly in the uplands and plateau region of central Mexico, mostly north of Mexico City, and the speakers of these languages would seem likely to have migrated northward from in or near what is now Oaxaca a matter of millennia ago. Apart from Oto-Manguean languages and a couple of language isolates (Huave, Tequistlatec), Oaxaca is also home to a large part of the Mixean branch of the Mixe-Zoquean language family, as well as some Zoqueans. Mixe-Zoquean has recently been grouped with the Mayan language family (David Mora Marin personal communication 2011), most members of which are located to the east, in Guatemala and Belize and in the Mexican states of Yucatan, Chiapas, and Tabasco. An outlying Mayan language, Huastec, is located in northern Veracruz and San Luis Potosí, and it may have been there for three thousand years or more. Mayan languages characteristically have glottalized stop consonants and affricates in addition to corresponding nonglottalized ones. Mayan languages have been linked to California Penutian languages, without widespread acceptance by linguists of the linkage. More recently a family called Toto-Zoquean has been proposed, to include the Totonacan languages and the Mixe-Zoquean languages (Brown et al. 2011). If this family has been appropriately posited, and if Mora Marin's evidence for including Mixe-Zoquean and Mayan languages in the same family is persuasive, then Totonacan languages will have to be included as well, suggesting that the family as a whole would extend along the Gulf Coast region (including some highlands) and east across the Isthmus to include the Mayan region.

The third main region of Mesoamerica, occupied when the Spaniards arrived primarily by Nahuatl-speaking Aztecs, is focused on the highland Valley of Mexico, including the states of Mexico and Tlaxcala, with additional speakers located in Morelos, Puebla, San Luis Potosí, and Guerrero. The language (or dialect group), existing in a number of dialects, has been called Aztec, Nahua, Nahuan, or Nahuatl (reflecting different emphases of time, grouping, or historical background). Nahuatl is in the Uto-Aztecan language family, and its closest relatives were the now extinct Pochutec language, once spoken close to the southernmost tip of Oaxaca, and the now moribund Pipil language (or Nawat), found today in El Salvador and at one time also in Guatemala, Honduras, Nicaragua, and perhaps as far south as the northern boundary of Panama (Paddock and Bernal 1970:67). Nahuatl has relatives that are more distant and more numerous north of Mesoamerica. In western and northwestern Mexico, there are Huichol, Cora, Tarahumara, Yaqui, and Mayo, for example. In the Great Basin of the western United States, linguistic relatives of Nahuatl include Papago, Hopi, Shoshone, Paiute, and Ute. The Aztecs, all speaking variants of the Nahua language, were relatively recent intruders in Mesoamerica, migrating from the north, some arriving in the Valley of Mexico perhaps as early as 800 or 900 CE. The latest to arrive and best-known of the seven separate tribes, according to legend, were the Mexica, who settled in the Chapúltepec region of today's Mexico City in about 1250 CE. According to tradition, in 1325 they resettled on the island of Tenochtitlán in Lake Texcoco. These were the people who developed an empire, conquering, exacting tribute, and trading with peoples far to the north and to the south of their island retreat. They were in Mesoamerica long enough to have borrowed many aspects of a partly shared worldview common to Mesoamericans preceding them, and they participated in its later development while also spreading rather widely some of their own words and concepts in the domains of religion, of place-names, and of names for a number of animals, plants, and items of trade. Much of this diffusion from Nahuatl, it must be suspected, came after 1521, when Nahuatl speakers accompanied their Spanish conquerors in expeditions of exploration and conquest in New Spain (Mexico and Central America, minus Panama, as well as a large chunk of what is now the United States west of the Mississippi) (see Brown 2011).

In the process of extracting communalities in the aspects of Mesoamerican worldviews that are associated with lightning, special efforts have been made to get information from representatives of each of these three regions and associated language families in order to deal with concepts that can indeed be generalized as Mesoamerican, and also to ascertain some of the more specific regional differences among the societies and their worldviews as they pertain to lightning and rain.

Language and Lightning in Mesoamerica

Many languages make distinctions among lightning, thunder, lightning bolt, sheet lightning, rain, and rainstorm that are similar to those in English, though not necessarily the same. Sometimes these languages share ambiguities in the application and understanding of the words for these concepts. Generally the audible (thunder) is distinguished from the visible (lightning) in nomenclature. Mesoamerican Spanish, for example, distinguishes the abstract phenomenon *relámpago* "lightning" from *trueno* "thunder" and both from a more concrete *rayo* "lightning bolt that strikes something," as well as from *centella* "(lightning) flash." On the other hand, Itzaj Mayan distinguishes *chaak* "thunder, thunderbolt, lightning bolt" (Hofling 1997:199) from *jatz'chaak* "lightning bolt" and *lem-lem* "lightning" (Hofling 1997:793). Totontepec Mixe distinguishes *vítsïk* "lightning, lightning bolt" from *añu* "thunder" (Schoenhals and Schoenhals 1965:264, 266, 280). In Colonial Tzeltal Mayan *tzantzewal* was "lightning," while *chahuk* meant both "thunder" and "thunderbolt/lightning bolt" (Ruz 1986). In Classical Nahuatl, the language of the Aztecs during the Spanish invasion in 1519, lightning was *mēyyō-tl* or *tlapetlānil-li*, the verb "to thunder" was *tlatopoca* or *tlatopōn(i)*, and lightning bolt was *huītecōc* (Karttunen 1983). The verb *tlatopōn(i)* is derived from the verb *topōn(i)* "make explosive noise, thunder, discharge firearm" (Karttunen 1983:247). A Classical Nahuatl word for lightning, *mēyyō-tl*, also meant "rays of light." Another Classical Nahuatl word for lightning, *tlapetlānil-li*, is derivationally related to *petlān(i)* "to scatter, glisten, reflect." One can understand something of Nahuatl *huītecōc* "*rayo*/lightning bolt" from Nahuatl *huītecu(i)* "to get whipped, to whip, beat someone."

A brief look at the composition and etymology of some other Mesoamerican words for lightning is also instructive. Yucatec Mayan *lelem* means "to flash (lightning)" (Kaufman 2003), but it also refers to a "machete-like object wielded in the heavens by the Chaaks" (Redfield and Villa Rojas 1934:116), rain deities visualized as old men riding their cloud horses. In addition to their machetes, modern versions of what must once have been jade axes producing thunder and lightning, these rain deities also carry gourds full of water from which come the rains. The term derives from Proto-Mayan **lem* "to shine, gleam," reflected in descendant languages, as, for example, in Pocomchí *lemow* "mirror," Tzeltal *lemlun* "to shine, glisten," Tojolabal *lemlun* "to burn," and Tojolabal *lemxi* "to clear up, brighten" (Kaufman 2003:472). These words and meanings can be compared with Huastec Mayan *lam* "mirror, glass" and *leey* "lightning" (Kaufman 2003:472).

Several members of the eastern branch of the Mayan language family have words for lightning—such as *kayopa*, *kayapa'*, *koyopa*, *kayumpa'*, *kaypa'* and similar terms—that are believed to have been borrowed from a Zoquean word that Kaufman (2003:473) reconstructs as **keyo-pa*. Eastern Mayan Pocomchi *kahoq* alternates with *kohoq* for a meaning of "thunder" and "lightning bolt/

thunderbolt." Western Mayan Tzotzil and Tzeltal have cognates *chauk* and *chahwuk*, respectively, meaning "thunder" and "lightning bolt." Kaufman (2003:489) reconstructs proto-Cholan **chahuk* "lightning, thunder," which is the same as /*chahuk*/ as spelled <cha-hu-ku> in the Cholan Classic Maya glyphic script, said to mean "lightning." The Yucatec Maya script variant from inscriptions on the Yucatan peninsula spell <cha-ki>, signaling the sound /*chahk*/, which is currently the name of the Yucatec rain deity (usually written Chaak). Sound correspondences indicate a Yucatecan borrowing of the name of this deity from a Cholan language, and the Yucatec day-name preserves the nonborrowed form Kawak (usually written Cauac) "rainstorm" as the name of the nineteenth day of the *veintena* (a twenty-day month). All of these cognates are reflexes of a Proto-Mayan word **kahoq* "thunder" according to Kaufman (2003:489), like contemporary Pocomchi *kahoq* "thunder, lightning bolt."

Tzeltal speakers refer to lightning with the word *tzantzel* (having no obvious etymology), and ball lightning (as well as meteors) with the word *paslam*, which has a Tzotzil cognate *poslom*, and cognates in a few other Mayan languages. Paslam, associated with fire, also names a Tzeltal monster with a head that is a ball of fire. Tzeltal *lem* "flashes of light" may be cognate with the root of Tzotzil *lebuel* "lightning," and Tzotzil *anhel* (a borrowing from Spanish) means not "angel," but "lightning" as well being an epithet of Yahval Balamil (literally "Earth Lord" or "Earth Owner"), the Tzotzil lightning deity. Quiche Mayan *anhel* means a "lightning bolt" or *rayo*.

Jacaltec Mayan (now called Popti') uses *k'uh* to mean "lightning bolt, thunderbolt, and thunder." In Classic Maya times the cognate form /*k'uuh*/ meant "god, divinity," and Kaufman reconstructs its etymon as proto-Mayan **k'uuh* "sun"(2003:458).

In the Mixe-Zoquean language family a reconstructed root **ne'm* with a posited meaning of "molest, bother" has reflexes in Mixean languages ranging from "make noise" and "molest" to "scandalize, shock" (Wichmann 1995:391). In Zoquean languages the reflexes include meanings of "to flash (lightning)," "to look angry," "to shine, glisten," and "to lick." Proto-Zoquean **nem* is given a meaning of "to burn," while its reflexes include "to flash (lightning)." Since Mixe-Zoquean languges have no /l/ phoneme, the similarity to the above mentioned Mayan meanings related to proto-Mayan **lem* "to shine, gleam" are one element of the evidence relating the Mixe-Zoquean and Mayan language families to one another.

Wichmann (1995:491) reconstructs proto-Oaxaca Mixean **witzuk* with a meaning of "lightning, lightning bolt," likely related to and perhaps derived from proto-Oaxaca Mixean **witz* "to carry off, to carry away/*llevar*." If indeed Mixe-Zoquean and Mayan language families are genetically related, then it seems at least distantly relevant that Huastec Mayan *witz* means "flower" (Huastec "candle" is literally "flower light"), while *witz* in other Mayan languages means "mountain, peak," which is where the lightning god lives and where lightning is more commonly found.

While Sayula Popoluca (a Mixean language) has *aniway* as the word for "lightning," meanings of cognates in other Mixean languages lead Wichmann to reconstruct proto-Mixean **'aniway* "thunder." Note that Lipp (1991:27–29) says the Mixe lightning god is named 'Ene• "Thunder." Totontepec Mixe *añu* is "thunder" and is likely related to *añ* "to prick, burn, sting." Wichmann (1995:248) reconstructs proto-Oaxaca Mixean **'an* for "hot," and proto-Mixean **'an* as "sting, burn."

In the Oto-Manguean language family, Chatino *xa³lwi²⁰* is "rayo/lightning bolt," which is composed of Chatino *xa³* "light" and *lwi²⁰* "brilliant" (Hilaria Cruz personal communication 2011; the numbers refer to tones in this tone language). Yatzachi Zapotec has *yi' gwzio'* "relámpago, rayo/lightning, lightning bolt" (Butler 1997:275). The *gwzio'* clearly comes from the name of the lightning deity, usually written as Cocijo and referenced as the god of thunder; the *yi'* means "light, firelight," so together the Yatzachi name for lightning is transparent. Yatzachi Zapotec "to thunder" is *chašjw chNiN* (**N** is **n** pronounced strongly and is in phonemic contrast with it). The *chašjw* means "explode, thunder," as does *chNiN* (Butler 1997).

For clarity of exposition some sixteen Mesoamerican cultural associations with lightning in the broad sense, inclusive of lightning, thunder, lightning bolt, and thunderstorm, are discussed here in an order that is intended to facilitate understanding the nature of some of these associations (see **Table** 2). No attempt has been made a priori to fit these associations to the Andean region, which has a different history and its own independent set of associations with lightning.

First comes the obvious association made between lightning and rain, which is often generalized to include other forms of freshwater, such as lakes, springs, and rivers. Much less obvious, and a point of departure from cultures of the Andes, is the special Mesoamerican association of lightning, rain, and clouds with caves. Lightning is also associated with communication with the Other World (whether the heavens or the underworld), highlighting ecstatic trance, shamanism, and sorcery, as well as kingship in manifesting such communication. Caves share these associations because they are seen in Mesoamerica as doorways to the Other World and as portals for communication with inhabitants of the Other World. The rain and lightning deity resides in his cave, which partakes of both worlds. The cave interface with this world is traversed by spirits of lightning bolts, shamans, deities, ancestors, and sometimes human co-essences (*naguals or tonos*)[4] that are said to be lightning. Lightning is also associated with warfare, with punishment for human transgressions, and perhaps

---

[4] For some, especially Nahuatl-speaking communities, and currently Spanish- but once Nahuatl-speaking communities, the *nagual* is not a co-essence but rather one kind of transforming witch—usually one that can, by performing various rituals including incantations, be transformed into a domesticated animal such as a pig, dog, sheep, or donkey (cf., e.g., Ingham 1986:118). In such communities, the co-essence is usually referred to in Spanish as a *tono* or *tona*.

TABLE 2
Mesoamerican Lightning Associations

1	Four Directions and Colors
2	Mountains and Caves
3	Shamans and Kings
4	Dwarf Helpers
5	Animal Familiars
6	Protection and Punishment
7	Fertility and Wealth
8	Bringer of Maize
9	Serpent, Eagle, and Jaguar
10	Twins and Triads
11	Thunderstones
12	Tobacco
13	Mushrooms
14	Frogs
15	Fish
16	Red

surprisingly with protection from danger. Where lightning bolts strike the earth they are believed to leave stone tools (of agriculture, hunting, or warfare), including jade axes, flint or chert blades, chalcedony, or worked obsidian. Lightning is also associated with serpents and indirectly with twins, and it participates in both a four-way directional organization and a triadic or three-way classification as well. In more than one region in Mesoamerica, a lightning deity is connected to crop fertility, riches, lizards, crocodiles, and occasionally snakes, and particularly with the nourishing riches derived from having split open a huge rock or mountain to reveal maize for the benefit of humans. Finally, lightning has somewhat more limited regional associations in Mesoamerica with tobacco, mushrooms, frogs, fish, and the color red. The associations are listed here in tabular form, in order of their discussion below. (To some extent the more significant associations come earlier than the others.)

Lightning Deities, Directions, and Colors

Lightning is associated with thunder and with rain for the obvious reason that they so frequently accompany one another; and thunder, lightning, and rain are all associated by indigenous Mesoamericans with a lightning-rain deity (or deities) responsible for sending these natural phenomena. It is to this deity that various forms of veneration, including sacrificial offerings, are directed rather than to the phenomena themselves. Usually the rain god is the same as the lightning god (see Marcus 1978:181), and the terms "lightning god," "lightning deity," "earth lord," and "rain deity" are here used for Mesoamerican deities more or less interchangeably.[5] These and several others are terms based on literal translations of diverse Mesoamerican languages' names for the rain god. The rain god is somewhat differently conceptualized in different Mesoamerican societies. What is surprising is that these rain/lightning deities in the different societies share as many features as they do. Using the different terms for what is clearly the same essential deity is a step toward recognizing the diversity within the unity.

The term "lightning deity" as used here refers to an anthropomorphic deity (that may or may not also have zoomorphic characteristics) who receives veneration from individuals in different societies and who is believed to use and control lightning. This minimal characterization references deities that both reward and punish humans; usually control such other weather phenomena as thunder, rain, and hail; often control wind; and are not infrequently considered the "master of animals." Such deities may appear to humans in animal form as well as in human form, and in some cases can appear as meteorological phenomena as well. It is the starkly defining criterion of lightning control that is most relevant to this male deity in Mesoamerica (who may have assistants, male or female). The other associations of this deity are noncriterial attributes. Because the lightning god is sometimes named Lightning in an indigenous language, sometimes Thunderer, and sometimes Rainstorm, he will from time to time also be referred to here as the rain god, or the thunder god, or the rain deity, the thunder deity, or lightning deity, and even the god of rain, god of lightning, or god of thunder. All groups discussed strongly associate lightning, thunder, and rainstorms with one another.

When placing the lightning deity within the contexts of specific cultural traditions, it becomes clear that while much is shared, there is also considerable variability in the details. Details are provided here primarily when necessary to document or explain the generalizations.

The most widely known of the Mesoamerican lightning-rain deities is the Aztec god Tlaloc.[6] He is often referred to as simply a rain deity, but clearly he is the rain-lightning deity (**Figure** 31; see color plate section), described shortly after the Spanish Invasion as "God of Rain, Thunder, and Lightning" (Durán 1971:154). He lives in mountain caves and on mountaintops and is the lord of water and earth (Heyden 1975:134). Characteristically sporting "goggle" eyes

---

[5] Although synonyms can be confusing, they are generally quite useful in conveying nuances of difference reflected in the specific instances of their use, and their use can limit what might otherwise be monotonous repetition. "Rain god," "rain deity," "lightning deity," "thunder god," and similar epithets are for the most part synonymous. When one is singled out in some cases it will be because due to the name or character of the deity being discussed, or to common usage by others regarding that society, that one epithet may be more appropriate than others. Similar reasoning underlies the more or less interchangeable use of "co-essence," "animal familiar," "animal spirit companion," *nagual*, *tono*, and so on.

[6] Where vowel length is specified as in Karttunen 1983, the name is given as Tlāloc in an orthography indicating a long **a** and the final **c** pronounced as a **k**. The Tlal- component in the name for this god of water, rain, and lightning refers to "earth" or "land."

and large teeth resembling jaguar canines (or serpent fangs, as some have suggested), early images of Tlaloc are found on pottery from two millennia ago in the southern Valley of Mexico (Miller and Taube 1993:166). He is also relatively frequently depicted in painted murals at Teotihuacan around the beginning of the current era and in diverse media at Postclassic Aztec sites, not to mention in various forms in Precolumbian Oaxaca and the Maya region, where central Mexican influence is indicated by virtue of these images. Tlaloc was thought by Postclassic inhabitants of the central highlands of Mexico to live in mountain caves that were "considered by local Nahuatl speakers to be miraculous treasure houses filled with wealth and prosperity" (Miller and Taube 1993:166).

The name of the Yucatec Maya rain-and-lightning deity is usually written as Chac, Chaak, or Chahk (**Figure** 32 and **Figure** 33). Other Mayan lightning-rain deities include Mopan Huitz-Hok "lightning deity" (Thompson 1930:57); Chuj Witz Ak'lik, literally "Mountain-Grassland" (Cruz Guzmán et al. 1980:116; cf. Thompson 1930:57); Huastec Muxi' Maam, the chief Maam "rain deity (literally 'Grandfather')" who rules the eastern quadrant (Alcorn 1984); Veracruz Huastec Junkil-aab, literally "he who accompanies rain" (Ariel de Vidas 2004:350);[7] and Tzotzil Chauc, also named Anhel (Thompson 1970:267; Vogt 1969). There are also the Jacaltec Komam K'uh, "Lightning Lord (literally 'Our Lord Grandfather')" (Spero 1991), and Chorti Chicchan, "rain deities that send rain, thunder, and lightning," manifested as giant snakes who occupy bodies of water and communicate with four giant rain gods bearing water jugs called Ah Patnar Uinicob (Thompson 1970:262–263). Additional Mayan rain and lightning deities include the Lacandón Menzabak, who lives in a cave next to Lake Menzabak and dispenses black powder for creating rain to Hahana K'u, Lacandón rain messengers that have been dispatched by Hachacyum, the creator god, who can strike the edge of his axe to create lightning bolts, which are personified as U Hahab K'u (Thompson 1970:265–266).[8]

Kekchi Mayan Yaluk, the chief Mam, an aged thunder deity (literally "Grandfather"), rules the western quadrant (Thompson 1930). Kekchi Mayan

---

[7] This literal rendition of the word by Ariel de Vidas has potential for revision in view of the proto-Cholan word **kil* "thunder" (Kaufman and Norman 1985:122), when the suffix -aab might be seen as constructing an abstract noun, or as a gentilic suffix, rather than as "rain" (Edmonson 1988:523, 326). We can also compare Veracruz Huastec *chook* "thunder, thunderbolt, lightning bolt" and Potosino Huastec *tzook* "thunderbolt, shotgun" with the name of the Yucatec lightning deity. Huastec *leey* means "lightning" and may be related to Yucatec *lemba* "lightning flash, lightning," possibly in some way also to sixteenth-century Isthmus Zapotec <laa> "lightning" (Urcid 1992:289), but not in any obvious way related to Chinese *lei* "thunder."

[8] Here we see a worldview variant in which the rain deity has been perpetuated in oral narrative as three different beings, each with its own duty with respect to lightning production. Menzabak is the cave-dwelling soot (or black powder) maker. Then there is the supreme Lacandón deity, Hachakyum, to whom Menzabak is subservient, to whom all traditional Lacandón speakers pray, and who wields lightning bolts. And there are the sacred rain makers, personified lightning bolts, that are messengers from Hachakyum to Menzabak telling him to make black powder, and these are called Hahana K'u (Thompson 1970:265–266).

FIGURE 32 Maya rain god and lightning deity Chaak, Chac, or Chajk, shown on page 16 of the Madrid Codex. Known as God B in the Maya screenfold books or codices, Chaak is usually depicted with a reptilian visage, a long, often down-curling snout, and two curved fangs projecting downward from the mouth (not obvious in this depiction). Chaak characteristically has a zoomorphic face, wears a (*Spondylus*) shell earflare, and is often associated with a serpent. (Public Domain PD-Art)

FIGURE 33 Classic Maya lightning deity Chaak is on the left side of this illustration, which is on a painted Late Classic Maya ceramic vessel. He holds an axe in one hand and a personified but unidentified object is grasped with the other. Serpent belly plates appear on his legs, and in front of him is an anthropomorphic jaguar baby, apparently ready to be sacrificed. (Photograph K521 © Justin Kerr, http://www.mayavase.com)

rain, lightning, and thunder deities are known as Mams, but they are also called Tzultacah "Mountain-Valley (Lords)" according to Thompson (1930:57), and *Cu:l Taq'a* in Josserand and Hopkins's orthography (2007:104). They are guardians of mountain peaks (Thompson 1970:275) and are sometimes thought of as two groups of deities in opposition to each other (Mam versus Tzultacah), the former being old and malevolent and more associated with thunder, the latter being younger and benevolent and more associated with the visual phenomenon of lightning (Thompson 1930:57–60; Wilson 1995:51–61). In fact, in many highland Mayan communities of Guatemala, "there is a contrast between vigorous and youthful lightning gods and gods of thunder, who tend to be aged gods of the earth and mountains. The aged thunder god is frequently referred to as Mam ['grandfather'], and it is probable that the ancient Maya god known as Pauahtun or God N is the Prehispanic form of this being" (Miller and Taube 1993:106) (Figure 34).[9]

Tequistlatecs, an intrusive people speaking a language usually attributed to the Hokan family, live just east of the southernmost portion of Oaxaca, Mexico, where they may have been for less than a millennium. They call their lightning god Labojna Guijuala "Owner (of the) Mountain" (Turner and Turner 1971: 206, 274) or Galrey Guijuala "Lord (of the) Mountain"(Turner and Turner 1971:225). Unlike Tequistlatec, most languages in the state of Oaxaca belong to the Oto-Manguean family, as does Chatino. The rain deity in one Chatino dialect is called Ho'o Ti'yu (Greenberg 1981:83). The linguistically related Zapotecs—there are at least five distinct Zapotec languages—also venerate a rain deity, the most important god of the Zapotec pantheon. In Coatec

[9] Note that Bassie-Sweet identifies God N of the Maya codices as Itzamnaaj of the Classic Mayas.

FIGURE 34 Late Classic unprovenanced Maya polychrome vase known as the Vase of Four Gods. On the left is the maize deity shown rising from Sustenance Mountain, which has been opened by a lightning strike from the lightning deities. To the right of the maize deity and facing his back are two Classic Maya Chaaks (rain and lightning deities). Michael Carrasco (personal communication) has tentatively identified these two as the young lightning deity standing in front of the old thunder deity. (Drawing by John Montgomery, © Foundation for the Advancement of Mesoamerican Studies, Inc., http://www.famsi.org)

Zapotec he is named Ngwzi', while in Miahuatlan Zapotec he is Mdi' (Rosemary Beam de Azcona personal communication 2011). In Mixtepec Zapotec he is called Ngùzì (Hunn 2008:47). In the southern Oaxaca mountain community of Lachixío he is called Bezihyyo' (Mark Sicoli personal communication 2011), but the Zapotec deity is more widely known as Cocijo (also Cociyo or Goxiu), which means "Lightning (bolt)" (Reko 1945:146). Dzahui ("Rain") is the name of the best-known Mixtec rain and lightning deity (**Figures** 35a–b; see color plate section) (sometimes spelled Savi, Zavi, or Sabi), while the Mixtecs of Nuyoo call him Ñu'un Savi (Joyce 2009:58; Monahan 1995:104–109; Ravicz 1965:80).[10] The Otomi deity associated with rain, lightning, and mountains is Hmü'ye (David Wright personal communication 2011).

Mixean speakers occupy a portion of southeastern Oaxaca on the western side of the Isthmus of Tehuantepec. Unlike Mayan rain deities, Mixe rain-lightning-thunder deities, named 'Ene· "Thunder," are not just quadripartite but many, each village having its own guardian 'Ene·. When addressed in ritual

[10] A mountain spirit named Taba Yucu, literally "dueño de la montaña," according to Josserand and Hopkins 2007:104), or Tabayuku, literally "spirit of the mountain," according to Ravicz (1965:39, 40), shares characteristics with the Mixtec rain deity (lives in cave on the mountain and is associated with water) (Ravicz (1965:79), but also has sirenlike qualities akin to those of the Yucatec and Lacandon Maya Xtabay, when appearing to men (Romney and Romney 1966:74).

contexts, however, 'Ene· is quadripartite (male, female, red, and white) (Lipp 1991:27–29). Mixe 'Ene·, "Thunder,"

> is an anthropomorphic deity who controls and causes lightning and the winds to produce rain for the growth of maize and other crops. When he puts on a special cape, rain falls... At the same time there are a multiplicity of 'Ene·, some of which live in the interior of the earth and are a source of money, *pohme:ñ*, and other forms of wealth, such as cattle, that can be obtained by petitioning within caves. (Lipp 1991:28)

A mountain lord who protects the personal guardian spirits that live within the mountains, the Mixe 'Ene·, "Thunder," whose calendrical epithet is Tum Hukpi· (One Root), is vigilant in protecting maize and also in protecting villages. "Each village has its own 'Ene· that protects it. When the 'Ene· of another village comes to do harm, he is fought off by the local 'Ene·" (Lipp 1991:28). Many scholars are of the opinion that Mixe speakers are descendants of the archaeologically known Olmec culture. We do not know the name of the Olmec lightning deity, but it is possible that he is depicted on a stone stela at La Venta (**Figure** 36).

Some Totonacs living in the Mexican states of Veracruz and Puebla have a rain deity named Tajin, which means "lightning" (Miller and Taube 1993:106), while others call the God of Lightning Aktsini', and still others refer to him as El Trueno Viejo or Jili' (Romero 1999:64–65). Such intralanguage variability highlights the generally closed corporate nature of some communities speaking the same language as others, while retaining or generating nomenclatural means of differentiating themselves from each other in creating their individual community identities.

Many indigenous Mesoamericans refer to the rain-lightning deity using the name of a saint, a practice engendered by the colonial Spanish friars whose suppression of indigenous religions in favor of Catholicism resulted in the adoption by many indigenous peoples of specific saints' names to reference their own traditional deities.[11] The specific saint varies from one group to

[11] The saints' names were not randomly applied. Often the basis of application related to specific characteristics of individual saints that could be likened to those of the traditional native deities, and/or to the birth date of the saint. For example, either the inverted cross symbol of the apostle San Pedro or the death of San Pedro Canisius on December 21 (the winter solstice) could explain why the lord of Hell is named San Pedro in Tzeltal Mayan tradition, and San Juan's June 21 birth date (the summer solstice) coincides with the summer rains, resulting in the chief Huastec Mayan rain deity, the Maam of the east, being called San Juan. Noteworthy too is the fact that the Huastec Maam of the west is called San Pedro, possibly because according to tradition San Pedro Nolasco died near the end of December, but more likely the Apostle San Pedro is the one referred to, a saint whose symbol is the inverted cross, recalling the fact that in the Maya script an inverted *ajaw* ("lord") glyph has a phonetic value of **la**, which in several Mayan languages sounds like the word *laj* "he died." A note on spelling: Huastec has vowel length, and in this language the thunder deity's name is spelled Maam. In some other Mayan languages it is spelled Mam.

FIGURE 36 Olmec stela 19 from La Venta. This probable depiction of an Olmec lightning deity, or perhaps a ruler masked and garbed as a lightning deity, shows him resting on a serpent (here a tropical rattlesnake), grasping an object with a large hole in it (recalling the unidentified object in the Maya lightning god's left hand in Figure 32), an object that here resembles in form the incense bags carried by priests or shamans in later times and that might have functional homologues in the water gourds or ollas carried by the lightning deity in Mesoamerican folklore and in the white-powder-filled rain-making amulets carried around the necks of Seri shamans. (Public Domain, Wikimedia Commons, Source Audrey and George Delange, User Madman2001)

another. San Juan may be the most frequently encountered saint referencing the rain deity in Mesoamerica. For example, Huastec Mayans call the chief rain deity San Juan (Alcorn 1984:58). Chols also call him San Juan (Cruz Guzman et al. 1980:116). Some Nahuatl speakers call the water deity who dispenses rain and resides in the Gulf of Mexico Sa Hua, "a name derived from the Spanish San Juan" (Sandstrom 1991:249; Sandstrom 2005:40). Chorti

Mayans, however, associate the lightning god with Santiago and say that his war shield has images of flames and fire (Girard 1962:251). Nahuatl speakers in Tlayacapan, Morelos, have syncretized San Juan the Baptist with Tlaloc the Aztec rain god (Ingham 1986:184). Nahuas of Sierra de Puebla refer to the rain god as Santiago (Govers 2006:218), while Totonacs of Coyutla call their rain deity Saint Bartholomew (Govers 2006:160). Totonacs of the Northern Highland region and along the coast call him San Juan Aktsini' (Govers 2006:160), and Totonacs of the Southern Highlands say he is San Miguel (Govers 2006:310), as do the Tzeltal Mayans of Pinola (Bassie-Sweet 2008:106; Hermitte 1964). Gonzalez (2001) identifies San Isidro as the syncretic version of the Zapotec Cocijo, god of rain, lightning, thunder, and hail. Mixtec speakers from the Mixteca Alta identify the rain god with San Marcos (Ravicz 1965:80).

Mesoamerican rain and lightning deities are often perceived as multiples of four, and they are linked to the four cardinal directions (or to the intercardinal quadrants), with their associated colors. Sometimes they are also associated with four forms of water (e.g., hail, rain, clouds, fog) (Foster 2005:167; Thompson 1970:269; López Austin 2002:37; Manca 1995:229).[12] As Foster puts it, regarding the Classic Lowland Maya rain deity,

> Generally regarded as a benevolent god, Chak has been continuously worshiped in Mesoamerica for a longer period than almost any other god. Like many aspects of the cosmos, Chak often assumes a quadripartite form. Each of the four aspects of the quadripartite Chak was linked to a cardinal direction and was associated with its own color.... Red Chak, related to the east.... White Chak, was associated with the north.... Black Chak corresponded to the west.... Yellow Chak was related to the south. [Red Chak] was particularly important to Maya rulers, who utilized the associated imagery of this aspect of Chak to denote their own rulership. (2005:167)

Thompson infers color-directional imagery related to lightning from a current Tzeltal Mayan dictionary and notes a contemporary Tzotzil Mayan municipality, San Bartolomé de los Llanos, "where lightning is referred to the four world directions and colors" (1970:269; 1934). López Austin records a Chol Mayan narrative in which four different colors of lightning (corresponding to different directions) are sent one after another to split open a stone mountain in which maize is hidden (2002:37; see also Manca 1995:229). In central Mexico, in order to help Quetzalcoatl in retrieving maize from Sustenance Mountain,

[12] Girard (1995:160), working with Chorti Mayans, discovered that the four corners are actually located in the intercardinal directions, or solsticial points, with red associated with the northeast corner, black at the northwest, and so on, green being in the center. Similar associations can be found with other Mayan groups (Girard 1995:67).

the gods "summoned the rain gods (Tlalocs) who represented the four cardinal directions and were associated with the directional colors blue, white, yellow, and red" (Morán 2007: 23; cf. Bierhorst 1992).

Mixe color associations with Thunder, the rain deity named 'Ene•, have already been noted. With the winds, which are part of 'Ene•, color-directional symbolism is even more pronounced (Lipp 1991:29–30). In some Mayan groups the lightning deity controls the winds, and in others the winds are separate. "In modern Chorti belief, the winds seem to have hived off from the rain gods, taking some of their functions and attributes" (Thompson 1970:270), and the Yucatec Mayan wind deities may have once been seen as "servants of the Chacs who later were merged with them" (Thompson 1970:271) and also have pronounced color-directional symbolism.

Concepts of rain and lightning are conflated in the nineteenth Yucatec Maya day-name, Cauac or *kawak*, which is given the traditional meaning of "rainstorm" and corresponds to the Aztec day-name Quiahuitl, meaning "rain," "thunderstorm," "fiery rain," or "lightning-storm." The Yucatec form Cauac can be traced etymologically back to proto-Mayan **kahoq*, meaning "lightning" and "thunder" (Kaufman and Norman 1984:117). The name of the Yucatec rain god Chaak, borrowed during Classic times from a Cholan Mayan language, derives from the same proto-Mayan form. This is additional evidence, through calendar day-names, of an ancient tradition that has continued to the present, demonstrating the close relationship among lightning, rain, thunder, and lightning bolts. The concepts of fire and great power, while peripheral to the main complex of associations here (i.e., they are noncriterial attributes), are nonetheless present in at least some measure with respect to lightning.

In short, the association of lightning with rain and rain deities brings out connections to four directions and four colors, going back to at least the time of the Classic Maya (Bassie-Sweet 2008:114, 251; Madsen 1955; Thompson 1934, 1970) (**Figure** 37; see color plate section). Fire and strength, as well as winds, can also be associated with lightning, directions, and colors. The connection of lightning deities to "four" extends to four water jugs (Brundage 1979: 70–72; Rosemary Beam de Azcona personal communication 2011) or water gourds (Alcorn 1984:59), and four kinds of water as well (Joyce 2009:58). In general the rain and lightning deities themselves, whether single, quadruple, or in some other multiple, have assistants who are usually dwarfs or "little people," as we shall see, and these are strongly connected to water, to the mountains, and to caves.

Lightning, Mountains, Caves, and Clouds

Mountains and caves, especially mountain caves, are closely associated with lightning, clouds, and rain, and it is through caves that lightning and the lightning-rain deity are connected to the Other World, because caves, like altars, are seen as doorways or conduits to the Other World. It may be observed that clouds, which are easy to associate with rain, and lightning, often tend to gather around mountain peaks. Lightning as well can be viewed in impressive display in the clouds of volcanic ash visible during eruptions seen on mountain peaks that are sometimes living volcanoes. Such displays are not soon forgotten (Shaw 1971:155). The frequency of lightning strikes, too, increases in the higher altitudes, as on mountains, and it is generally greater with closer proximity to the equator, so lightning's association with mountain peaks in traditional lore is to be expected on the basis of simple observations.

It is sometimes difficult for Westerners to comprehend indigenous convictions that the lightning deity guardian of the mountain, approachable through caves, sends clouds, rain, and lightning onto the earth from caves. However, the indigenous point of view on this association is widespread and well documented (Brady and Prufer 2005:110; Chuchiak 2000:267; Andrews 1970; Thompson 1970; Vogt and Stuart 2005; Gossen 1972:136–137; Morris 1987:105–119; Heyden 1975:134; Lipp 1991:49; Fitzsimmons 2005:110, 1997:1–3; Bassie-Sweet 1991, 2008; Earle 2008). Even observant Westerners have on occasion noticed an empirical basis for the connection (**Figure** 38; see color plate section). For example, in the highlands of Chiapas, Vogt (1969:387) maintains that actual observations of clouds from the lowlands climbing up and pouring over highland ridges strongly support indigenous perceptions of cave origins for clouds, rain, and lightning. Bassie concurs and elegantly explains the indigenous perceptions.

> The view that mountains, caves and the deities that inhabit them were the source of wind, lightning and water was based on visual observations. The numerous springs and streams which percolate up from the interior of a mountain, either through a cave opening or the ground itself, visually demonstrate that mountains are a primary source of water. Rising warm air forms clouds on the tops and slopes of mountains, leading to the conclusion that clouds and rain come from there. Mist invariably forms at the mouths of caves, reinforcing this belief. The cool breezes that blow from the mouths of caves help explain the belief that wind originates from there as well. During thunderstorms, the tops of mountains flash with lightning, leaving the impression that lightning comes from these sacred summits. (2002:6)

According to Chuchiak (2000:267), even on the lowland peninsula of Yucatan, where there are no mountains, the Maya of colonial times located the Chaaks (rain gods) and Pawatuns (wind gods) in caves and cenotes (sinkholes in the local karst geography), and lightning's predilection for striking the highest object in a plain results in more frequent strikes atop the pyramid temples. In the lowland sites of the Classic period Maya, there is considerable evidence of rain ceremonies to be found in caves. Such associations are found from the Late Preclassic at La

FIGURE 5 Lightning bolts or *rayos* striking the mountain peaks. Lightning bolts that strike the earth are *rayos* and have both natural and symbolic associations to mountain summits, referred to as *apu* or mountain lords. Mountain summits are often near sources of streams and thus have a symbolic association to the channeling of fluids, the male generating principle.

FIGURE 13 The Cordillera Blanca in central Peru includes some of the highest mountain summits in the world. Mountain summits were called *wamani* or *apu* and have masculine associations and are commonly associated with cloud-to-cloud lightning or *relámpago*. Crosses placed on mountain summits and passes symbolized both cloud-to-cloud or horizontal (*relámpago*) and vertical lightning bolts (*rayo*). (Photo by Harold Hietala)

FIGURE 17 Pumapungo or "Door of the Puma" at Tomebamba, Azuay Province, Ecuador. Tupac Yupanqui commissioned the Temple of Pumapungo in honor of the Creator, Viracocha ("foam of the lake" [Quichua]). The Cañari and surrounding indigenous cultures venerated Viracocha long before Inca expansion. Artificial-stone-faced terraces demarcate the temple Pumapungo. Chroniclers and Spanish clerics later associated such architectural modifications with lightning *huacas*. (Photo by John E. Staller)

FIGURE 18 Cathedral of the Immaculate Conception, Cuenca, Azuay Province, Ecuador. The Catedral Metropolitana de la Inmaculada Concepción represents the spiritual heart of the city of Cuenca. Its three blue-and-white domes have important cultural and historical significance to its citizens, as do the religious beliefs surrounding the Immaculate Conception. The cathedral was a monumental work of faith, renowned for its Neo-Gothic style. (Wikipedia Commons)

FIGURE 20A Sol de Oro was initially believed to represent a lunar calendar looted from a tomb in Patecte, near Chordeleg, in southern highland Ecuador. Museum officials in Guayaquil affirm that the gold mask was looted from an ancient tomb at Chuncari, near Sigsig. The emanating serpents and their undulating bodies are symbolic of Illapa. In their origin myth, the Cañari were transformed from snakes into humans that emerged from the surrounding ponds and lagoons into the present creation or world cycle. Such concepts may have been represented in this gold mask (Gold-copper alloy, second or third century AD, Collection of Museo Antropológico del Banco Central del Ecuador, Guayaquil; photo by John E. Staller)

FIGURE 20B Neckless *olla* with red slip and fine-line incision from Cerro Narrío in layers dated to c. 1400–1200 BCE. Cerro Narrío is an ancient hilltop *huaca* near Cañar, Ecuador. The red slip is associated with *Spondylus*, the jagged double-line or step motif with serpents and symbolic representations of lightning or Illapa. The Inca and other cultures used the vertical jagged double-line symbol to refer to *rayos* or bolts. (Courtesy of the Field Museum Collections; photo by John E. Staller)

FIGURE 21A Thorny oyster was called *mullu* in the Andes and traditionally served as ritual offerings or *pagos*, and funerary offerings associated with lightning and its related *huacas*. *Spondylus* shell was important throughout pre-Columbian Central and South America and highly valued for its red color, which symbolized blood and female sex (*vagina dentata*) and the fertility and fecundity of the earth and perceived as precious in both Central and South America. Thorny oyster (*Spondylus princeps princeps* L.) is represented above, while the larger samples represent the southernmost species of this genus *Spondylus calcifer*, which ranges solely along southern coastal Ecuador and far northern Peru. All were excavated from the ceremonial center of La Emerenciana, El Oro Province, Ecuador. (Photo by John E. Staller)

FIGURE 21B The *Strombus* Conch (*Strombus galeatus* L.) is the heaviest-shelled of this genus of gastropod. Its colors are variegated brown-and-white bands, or orange-yellow blotches. *Strombus* was crafted into status-related objects or modified to function as wind instruments or *pututus* (trumpets). It begins to appear along with *Spondylus* as offerings as early as the Formative and Preceramic periods in Ecuador and Peru. These were ritual offerings from the ceremonial center of La Emerenciana. These marine shell species were traded early on in the context of long-distance exchange. (Photo by John E. Staller)

FIGURE 27 Rainbow preceding a storm in the sacred Urubamba valley, north of Cuzco. When trade winds blow storms up the eastern cordillera, rainbows sometimes appear before a storm approaches. This phenomenon has important implications for indigenous Andean folklore and legend surrounding lightning, particularly beliefs associated with the cat of the *apu*.

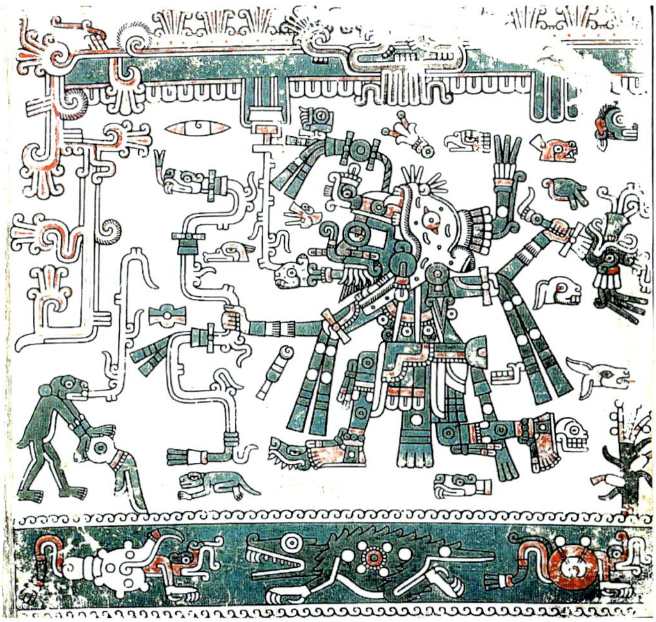

FIGURE 31 Aztec Tlaloc from Codex Laud, folio 2. Tlaloc, the Aztec lightning/rain deity, is striding with a lightning serpent in his right hand and a stone axe in his left. (Public Domain PD-Art)

FIGURE 35A Mixtec rain/lightning deity, Dzahui, represented on a Mixtec style Post-Classic polychrome ceramic vessel from El Chanal, Colima, Mexico. Vessel is in the Kimbell Art Museum, Fort Worth, Texas. (Public Domain—Courtesy of User:FA-2010 PD Wikimedia Commons)

FIGURE 35B Helpful (left) and destructive (right) aspects of the central Mexican rain deity, depicted on page 27 of the Codex Borgia. The image on the left shows personified maize ears growing happily, watered and blessed by this rain deity. On the right, the maize crop is being destroyed by the same deity. The facsimile edition of Codex Borgia is online at Famsi.org. (Public Domain PD-Art)

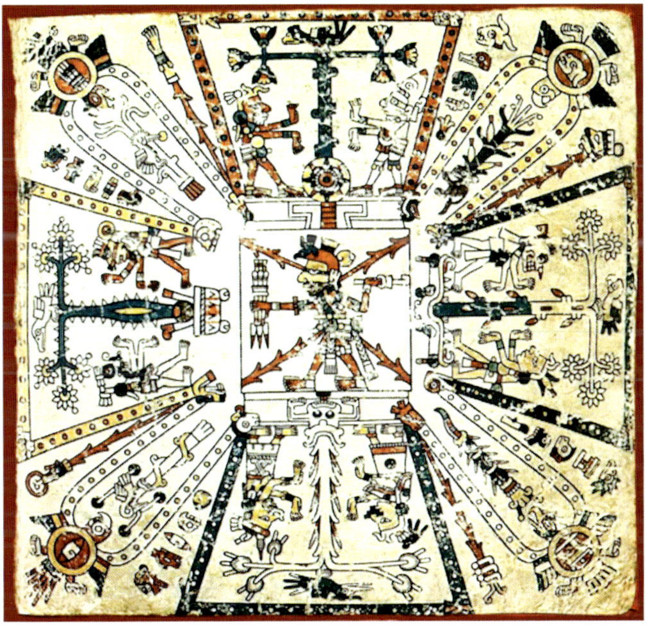

FIGURE 37 Color-directional symbolism is illustrated on page 1 of the Fejérváry-Mayer Codex, created in Mexico in the fifteenth century. The top arm of the directional cross is red and represents the east; north is yellow, and so on. Each cardinal direction has its own specific tree, bird, and pair of attendant deities known as lords of the night. Of the nine lords of the night represented, the one in the center is probably Xiuhtecuhtli, who is both Lord of Fire and Lord of the Year. (Public Domain, Wikimedia Commons PD-Art)

FIGURE 38 Chalcatzingo Monument 9, representing a cave entrance zoomorphically depicted as the mouth of the mountain lord/lightning deity's animal familiar or co-essence, which in some cases is a serpent, lizard, or crocodile, and in others perhaps a jaguar. This mouth is in the shape of a cross and symbolizes not only a cave, but also a portal between cosmic realms or worlds. (Photo K3116 © Justin Kerr www.mayavase.com)

FIGURE 41 A dwarf depicted on a Classic Maya vase is shown playing a significant role in an apparent discussion or face-off between groups. (Photo K632 © Justin Kerr; www.mayavase.com)

FIGURE 45 Ch'a Chaak ceremony with boys representing frogs tied to the four legs of the altar. They make frog sounds to heighten the symbolic effect. This diorama is in the National Museum of Anthropology in Mexico City. (Photo courtesy of Thomas F. Aleto)

Palita Cave in the Guatemalan Peten region, where "inside the low overhanging shelter was a 'life size' masonry effigy of the rain god Chaak, seated on a throne and covered with modeled stucco" (Vogt and Stuart 2005:163). Cave associations with rain and with rain deities are also well established for Huastec Mayans. Like most Mayans, today's Tzotzil speakers believe that "all forms of precipitation, including clouds, lightning, and thunder" (Gossen 1972:136–137) are controlled by the Earth Owners [rain-lightning deities] who dwell in mountain caves.

Vogt and Stuart supply some additional context when enumerating what has been learned about caves in highland Chiapas, and the narratives local Mayans tell about them.

> The mountain caves are portals to the Underworld, and the source of lightning, thunder, clouds, and rain, which are controlled by the mountain gods who live in these caves with their virginal daughters. These daughters spend their time fluffing cotton, which will be transformed into rain clouds by the fertilizing lightning bolt of Anhel. The rain-laden clouds then issue forth from the mouths of caves, beginning the rainy season and bringing life to the land. Sometimes the lightning is used to punish living members who have misbehaved, but more often it is employed to bring the rainfall needed for the maize fields. (2005:177; internal citations omitted)

From the Popol Vuh and other Mayan narratives, Bassie-Sweet reconstructs a sacred landscape in which a quadrilateral world with a mountain at each of four sides represents the house of the creator grandfather, identified in the Popol Vuh as Xpiyacoc. She goes on to identify this being as the turtle associated God N of the Postclassic Maya codices and also Itzamnaaj of the Classic Maya inscriptions (2008:64). An earth deity associated with four mountains and called grandfather recalls Mountain Lord or Earth Owner of today, living in mountain caves and sending lightning as reward and punishment. It is his manifestations, helpers, and tools that constitute the variety of lightning and related weather phenomena (thunder, meteors, rain, winds, whirlwinds, clouds, and hail) seen by inhabitants of contemporary Mayan communities as one aspect or another of the rain god or lightning deity.

Looking to the Classic Maya inscriptions for parallels of contemporary deities is always a somewhat tenuous project. At this stage it looks like the lightning associated deities of the Classic Maya of the current era's first millennium may have had multiple personae, avatars, or manifestations of what some contemporary peoples identify as a single deity. For example there were different moon goddesses, and multiple maize deities. Even today in some Mayan societies we have evidence of older rain gods and younger rain gods (e.g., Kekchi communities), and Lacandóns implicate three different kinds of deities in the delivery of lightning and rain. It seems entirely possible that God N/Itzamnaaj is a lineal ancestor in the narrative leading to the

Tzotzil Anhel (Earth Owner), who is now their lightning deity and rain god, living in the mountain caves and sending lightning and rain where he pleases.

In the Aztec- (Nahuatl-)speaking central Mexican highlands, Tlaloc was the earth lord, identified by all as the rain deity. Tlaloc lived in caves and also on mountaintops (Heyden 1975:134). The Mixes of Oaxaca also connect the rain-lightning deity with caves, as noted by Lipp (1991:48–49). In pre-Hispanic Oaxaca, "the principal deity of the earliest people at Monte Albán was the jaguar/baby-faced god of rain and lightning, who lived in a cave and who, according to Burgoa, was the ancestor of the Zapotec rain-lightning god, Cocijo" (Fitzsimmons 2005:110; internal citations omitted) (Figure 39). Numerous additional citations demonstrating connections between caves and the rain-lightning deity in various Oaxacan language groups, such as Mazatecs, Ixcatecs, Mixtecs, Mixes, and Zapotecs, Chinantecs, and Oaxaca Chontals, are provided by Fitzsimmons (1997:1–3; 2005:110). In fact lightning, the lightning deity, and caves are integrally related throughout Mesoamerica, as has also been discussed by various other Mesoamericanist scholars (Brady and Prufer 2005; Bassie-Sweet 1991, 2008; Earle 2008; Thompson 1970).

To summarize, all over Mesoamerica, from Formative times to the present, lightning, clouds, rain, and the deity associated with them have all been closely connected to caves. The rain-lightning deity, sometimes called Earth Lord or Mountain Valley,[13] or just Anhel—a term borrowed from Spanish Catholicism but with a meaning quite different for the borrowers from that which an angel had for the Spaniards—inhabits caves, according to multiple Mesoamerican traditions, and is venerated in caves and near their entrances. Sacrifices are made there and favors asked. It is easy to understand how caves may be seen as entrances or conduits to the Other World (usually the Underworld). It is precisely in this way that they are, and have long been, viewed by indigenous Mesoamericans. Lightning, which is strongly associated with caves and with veneration of the lightning deity, conducted on mountain peaks and in caves, is also associated with shamanism, ecstatic trance, and sorcery. To the extent that a form of what has been called "state shamanism" (Reilly 1989; Chang 1983) was one of the foundations of kingship in Mesoamerica, lightning and connected concepts also can be associated securely with kingship, even without the additional evidence presented below.

---

[13] Mountain Valley in this context represents a *difracismo*, typical of Mayan languages (as well as Nahuatl), the meaning of which in this case is "Earth." A *difracismo* is a figure of speech found in several indigenous Mesoamerican languages in which two separate words are paired together to form a single metonymic compound. In Tzeltal, for example, the word for "plant," *te'ak'*, is composed of *te'* "tree" and *ak'* "vine"; if paired, whether immediately juxtaposed or separate components of a couplet, *te'* and *ak'* refer to "plant." Similarly Tzeltal *chanbalam* "animal" is composed of the words for "snake" and "jaguar."

FIGURE 39 A carved ceremonial or votive jade axe well known as the Kunz Axe, reputed to have been found in Oaxaca, depicts what has been identified as the Olmec baby-faced jaguar rain deity. (Wikimedia Commons license, Public Domain, Photo by Madman2001)

Lightning, Shamanism, and Kingship

The terms "shaman" and "shamanism" have been hotly debated in anthropology, and while "shaman" names a specific type of Siberian healer, it is sometimes used as a synonym for healer, medicine man, curer, herbalist, or diviner in any non-Western society. This broad formulation can result in making non-Western healers seem exotic and different, which potentially has subjective evaluative connotations that are undesirable. In fact there is some substance to claims that the words "shaman" and "shamanism" have been overused and misused at times, leading to misunderstandings (Klein 1999), and even the argument that they should not be used outside an appropriate Siberian context cannot be dismissed out of hand (Kehoe 2000). However, the long and useful employ of the term "shaman" in Mesoamerican and other non-Siberian literature despite its association in some cases with ethnocentric perspectives, suggests both that it can be a useful term and that it is necessary to carefully delimit what we mean by "shaman" and the associated word "shamanism." Tedlock (1992:47–84) has a useful discussion of the term "shaman" as used in Mesoamerica, and following a brief note on general specialists in wholeness or wellness, I will delineate how the term will be used in this chapter.

A shaman is not just any kind of curer but, rather, a specific if not uncommon type among present-day indigenous populations in Mesoamerica.[14] A wide variety of different kinds of healers, curers, diviners, and related specialists can be found in Mesoamerican societies. They can be recruited by various means, make their diagnoses with different instrumentalities, and effect cures with distinct, diverse, and multiple methods. They may be accorded different statuses in different home communities. Some cure highly specific illnesses, diseases, or health problems, like the Aztec *tepopoque*, whose specialization is the cure of *aire de cuevas* (cave air) afflicting individuals who have offended the dwarf assistants of the lightning deity (Madsen 1955), or the Yucatec Maya bonesetters, who just deal with broken or displaced bones (Paul and McMahon 2001); others might diagnose and cure a variety of ailments, while also performing midwifery. Some use eggs to diagnose illness or in the curing process (as, for example, in rubbing over an afflicted area in order to draw out the disease, which can then be disposed of). Some healers diagnose by means of flashes of what they refer to as "lightning" coursing through parts of their bodies, especially their legs. Others diagnose illness with different divinatory techniques, such as animal sacrifice, gripping and interpreting the patient's pulse, or ecstatic trance, with or without the aid of alcohol and/or psychoactive botanicals such as peyote, mushrooms, datura, *Salvia divinorum*, or tobacco (Wilbert 1987). Some healers cure illness with prayer, and/or

---

[14] There is no reason to suppose a priori that shamans do not exist in urban contexts in contemporary civilizations, and in fact there is some evidence that they do.

with sacrifices of blood, animals, or food. Some prescribe or prepare medicine bundles, while others suck out purportedly intrusive materials, blow a mist of saliva on afflicted people or their parts, or sweep away the perceived health problem with hawk plumes, and/or the burning of incense. Some use herbs (and sometimes other natural products) for patient ingestion. The point is that each society is likely to have its own set of named categories for specialists that diagnose and cure conditions that threaten the life and wholeness of individuals and communities, and these are not necessarily directly comparable (or isomorphic) with the categories and the specific practices that are found in other societies. At the same time some of these healers fulfill the conditions that we believe justifies using the word "shaman."

A "shaman" here designates a specific kind of healer, one that attains a state of ecstatic trance, dissociated from his or her present reality, in order to accomplish such ends as diagnosing illness, prescribing cures, finding lost objects, predicting future events, and influencing the weather. These individuals interpret their entry into an alternative state of consciousness as their soul journeying to another realm (e.g., the Underworld or the Heavens), thus accessing another plane of existence (see, e.g., Madsen 1955). This, then, is a kind of "ecstatic specialist" who can control the travels of his own soul and who "specializes in the knowledge and care of the souls of others" (Sullivan 1988:390). This generalized characterization with limited criteria allows us then to specify what we mean in using the term "shaman." There are a number of other regionally specifiable characteristics that accrue to shamans, and when relevant these are referenced.

The word "shamanism" is used in conjunction with a particular and simply qualified worldview and the activities of healer practitioners adhering to that worldview. It is a worldview in which a three-tiered cosmos (comprising the Underworld, Middle World, and Upper World) and all within it is assumed to be animate, and animated by a spiritual essence or life force; a cosmovision in which individual body and soul are separable entities under specifiable conditions that are capable of rejoining (see Markman and Markman 1989:102). The souls of shamans are capable according to their own interpretation, especially during ecstatic dissociation or trance, of leaving their body and visiting an alternative reality that can be called the Other World in lieu of specifying Underworld (or World Below) or Upper World (or Celestial Realm).

Kings, rulers during part of Mesoamerica's history, were powerful individuals authorized for governance of the state whose public activities apparently included some symbolic actions characteristic of Mesoamerican shamans (see Tedlock 1992:47–85). They clearly believed themselves responsible for much of the general welfare of the community or the polity being ruled. To judge from Classic iconography, the king served as a protector of his people through symbolic reference to warfare, so on occasion he was depicted in Classic period images carved on stone as a warrior, shown with weapons in hand (**Figure** 40). He was also a symbolic provider for his people, ensuring abundant crops and good weather, so

FIGURE 40 Late Preclassic Maya ruler portrayed as warrior and protector of his people, showing his weapon, which is edged with obsidian flakes. (Schele drawing—© Drawing/photograph by Linda Schele, © David Schele, courtesy Foundation for the Advancement of Mesoamerican Studies, Inc., http://www.famsi.org)

he was also sometimes depicted in images as a personification of the maize deity. He accomplished the protection and provision of his subjects, according to reputable Mayanists, by operating as a symbolic shaman, assumed by the populace to be able to traverse the barrier between worlds in order to contact the Other World and negotiate with deities, spirits, and ancestors that to this day are thought by Mayan descendants of the Classic Mayas to be capable of rewarding or punishing people of this world (Freidel et al. 1993; Reilly 1994; Looper 2003).

The ruler's display of his power and legitimacy to rule was in part accomplished by relating himself through imagery to lightning and the lightning deity (Looper 2003, 104, 133; Joyce 2009:215; Tedlock 1996:224; Freidel et al. 1993:196, 197, 200, 445; Tate 1992:94; Taube 1992a; Baudez 1992; Coggins 1988). For example, a Zapotec "monument discussed by Urcid (2005:141) shows a personage dressed as the rain deity handing a symbolic lightning bolt to a successor, perhaps his son" (Joyce 2009:215). It is significant, as will be seen below, that an important symbol of the Maya ruler's authority in Classic times was a two-headed "serpent bar" and that the serpent is for Mesoamericans in general an important symbol of lightning and of the lightning deity.

Returning to the shamans of more recent times, there is considerable evidence to indicate the importance of lightning to shamans and shamanism in Mesoamerica. An excellent example of a modern-day shaman is Don Soltero of the Nahuatl speaking town of Tecospa, a few miles south of Mexico City. During a storm lightning struck his home and he was knocked unconscious, reviving only slowly. While he was unconscious rain dwarfs kidnapped his soul and brought him to their cave in order to convince him to become a curer. He was reluctant, so they beat him as a form of persuasion. It took six months of such persuasion during weekly bouts of unconsciousness (or dissociative states), but Don Soltero became thus a shaman, referred to by the local villagers as a *tepopoque* or in Spanish a *curandero de aire* (Madsen 1955). In some other Nahuatl-speaking communities shamans surviving lightning strikes are called *graniceros* (or *granizeros*), and they too are seen to have an obligation to serve as rainmakers, maintaining control over various weather phenomena (Albores and Broda 1997; Lorente Fernandez 2009:1; Iwaniszewski 2001:405).

To the southeast of Mexico City, on the slopes of twin volcanoes, cave ceremonies of rainmakers also attest to the relationship between lightning and these shamans.

> Guillermo Bonfil (1968) described *granizero* rites in the caves at Iztaccihuatl and Popocatépetl. Supposedly there are schools to train *granizeros* [rainmakers] in these mountains.... In order to become a 'member and attend the cave ceremonies, a person must have been "called from above"—that is, designated to work on earth with the supernatural powers that manage rain, thunder, lightning and the climate. But only those who have been struck by lightning are eligible

> for this important task. If they survive the baptism by lightning, they cannot deny their destiny and must join the organization. (Brady & Prufer 2005:24–25; internal citations omitted)

In Tlayacapan, Morelos, a mostly Spanish-speaking village of Nahuatl heritage, rain spirits (*ahuaque, aires del tiempo*) in groups corresponding to local mountain shrines are said to do battle with each other using lightning bolts fashioned from gunpowder,[15] and lightning bolts are thought to leave *aires* that cause a special kind of illness in the ground where they have struck (e.g., insomnia, loss of appetite and spirit encounters). Here too there are *graniceros*, and they are charged with treating those who have encountered such *aires* or who have been struck by lightning.

> People who have been struck by lightning and survive the ordeal are understood to have an obligation to serve the weather spirits. Being struck by lightning is thought to give one a gift for controlling bad weather and for curing *aire* illness and other afflictions. Those who agree to serve in this fashion are called *sirvientes* (servants), *graniceros*, or *trabajadores temporales* (weather workers).... After making a commitment the person is inducted into the local group of graniceros in a ceremony called a "coronation," which graniceros from the surrounding region attend. The rain spirits are summoned with rockets and offerings of white flowers, chicken mole, fruit, bread, tobacco, alcohol, candles, and incense. (Ingham 1986:157)

Lipp gives an account of how different kinds of Mixe curers come to practice their crafts. One kind is approached in a dream by the lightning-rain deity with whom he makes a pact and receives instructions on how to deliver sacrificial offerings as well as on how to employ the appropriate curing techniques. Subsequently the shaman enters a trance state, often triggered by thunder and lightning, during which instructions are given regarding a patient's condition and cure. Sometimes in a trance the shaman will fly to the home of the lightning deity, high up on the mountain (Lipp 1991:152). On the other side of the curing coin, "indications of a sorcery attack are dreams of lightning striking near the house" (Lipp 1991:153).

In Veracruz Sierra Popoluca traditions certify that transforming shamans (here called *naguals*) "send forth lightning bolt men (*tota'huan*), who have the task of protecting the community and the harvest and whose patron is Saint

[15] The relationship between gunpowder, thunder, lightning, and rain is symbolically recognized in various other ways as well in central Mexico. The villagers of Tlayacapan use gunpowder-filled exploding skyrockets to ward off approaching hailstorms (Ingham 1986:157), and in Tepoztlán, when rain is desired, offerings of pulque and gunpowder are made in certain caves (Ingham 1986:197). Recall here the Lacandón Menzabac "Black Powder Maker," whose black powder is sprinkled on the clouds, releasing the rain from them.

Michael" (Báez-Jorge and Báez Galván 2005:152). The rain dwarfs (*chaneques*) in the Popoluca worldview, live in caves, rivers, and woods. Water *chaneques* are conceptualized as crocodiles that transform themselves into other animals so as to punish persons who don't adhere properly to social norms, other *chaneques* become serpents to punish humans (Báez-Jorge and Báez Galvan 2005:151).

In Oaxaca Mixtec rain shamans (*tenuvi*) have the power of a lightning bolt, with which they can cause deadly harm to one another (Monaghan 1995:349), and a co-essence kind of connection to powerful animals or to a lightning bolt (Monaghan 1995:173, 348).

Near Tila in Chiapas, Mexico, a Chol Mayan novice shaman will go to a cave, where he "must visit Uran, the Lord of the Mountains, the Lightning, and the Animals. The animals are ferocious until they are calmed by Uran after he receives two bottles of *aguardiente* and grants the novice shaman's request that he be taught how to cure illness" (Vogt and Stuart 2005:176).[16]

Quiché Mayan diviners of Momostenango are informed by "lightning" in their blood and muscles (*coyopa* in the Momostenango dialect means both blood lightning and external lightning). "Momostecans whose blood speaks experience the movements of *coyopa* many times a day, whether or not they happen to be considering a question about past or future events at the time" (Tedlock 1992:138). Quiché oral narratives also make reference to a being called both White Dwarf and Red Dwarf who carries a lightning striking stone hatchet and whose epithets clearly identify him as a Mayan version of the central Mexican rain dwarf. He is seen as responsible for the "lightning" in the blood of diviners, where his "child" (the lightning) lives to this day. This C'oxol (dwarf) in various accounts is portrayed as a gamekeeper or guardian of wild animals, as a small boy, and as the spirit of the mountain. He also has a breast-plate and sandals the color of blood (Tedlock 1992:148–150).

The shamanic linkage to lightning leads into another connection with kingship and earlier times in the lowland region of the Classic Maya. In the Popol Vuh Red Dwarf is an assistant to One Leg, a lightning-rain deity. "His lowland Maya counterpart, variously labeled God K, GII, or 'the manikin scepter' by iconographers, sometimes takes the form of a fire-striking or a lightning-striking axe, with the head of the axe sticking out of an obsidian mirror on his forehead" (Tedlock 1996:224; Freidel et al. 1993:200). God K as the manikin scepter has a serpent for a leg and serves the ruler holding him as a symbol of power and authority. He is shown on occasions of accession and bloodletting (Tate 1992:94), and was thus employed as a symbol of kingly lineage and succession. God K of the Postclassic screenfold books is K'awil of Classic times, with a serpent leg, and a serpent co-essence. His name, K'abwil,

[16] Incidentally, *aguardiente* is rum, sometimes called "white lightning"; the Spanish term is derived from the words *agua* "water" and *ardiente* "burning."

means "saint's statue and spirit" in contemporary highland Guatemala (Freidel et al. 1993:196, 197). In contemporary Yucatec, K'awil means "sustenance" (Freidel et al. 1993:194), recalling the lightning god's service in breaking open Sustenance Mountain according to Mesoamerican traditions, making maize available to humans. Thus, in addition to being a rain deity and associated with lightning (Taube 1992a; Freidel et al. 1993:445), K'awil is related to kingly authority, sustenance, and serpents. The latter two categories will be reprised below.

Independently Looper (2003) uses similar evidence to link lightning to kingship in Classic times but emphasizes lightning's power. "Further, God K has a close association with mirrors, one of the most ancient Mesoamerican symbols of the portal to the spirit realm, out of which the gods and ancestors are reborn. Finally, as a scepter (actually a personified axe) the being represents the power of lightning, wielded by the king. . . . Together these elements develop a persona for the ruler as lightning deity" (2003:104; internal citations omitted). Furthermore "as a lightning axe, the scepter relates to the king's role as a rainmaker" (Looper 2003:133).

To summarize, it should be clear that lightning and the lightning deity are closely linked to shamanism and trance in several ways, a phenomenon that sporadic reports indicate continues through Central America down into South America. Then, too, shamanic components of kingship, as exposed in particular through studies of Classic Maya rulers, provide reasons for the important association of lightning with these rulers. Sorcery is associated with lightning precisely because shamanism is so associated, as similar powers and associations apply to both in Mesoamerica.

The Lightning Deity's Dwarf Helpers

Bassie-Sweet, using data from various Mayan language narratives, says that "the portrayal of thunderbolts as small children or dwarfs is found across highland Guatemala" (2008:105), and it is likely that this portrayal pertains to the lightning deity's assistants, who clearly might also be characterized as manifestations of the lightning deity. Elsewhere in Mesoamerica, the rain-lightning deities often have multiple small assistants, usually dwarfs or very small individuals (often described as about a foot and a half, or as much as three feet, tall); individuals who are generally connected with water in the form of springs, pools, lakes, rivers, and creeks (Bassie-Sweet 2008:114, 251; Huber 1990; Madsen 1955; Chevalier and Sanchez 2003:126). They usually live in caves, as do the Nahua rain dwarfs, who are known as *yeyecatl, aguatoton, aires, chaneques, tamatinime,* or *enanitos* (Madsen 1955). These assistants are powerful and either can be helpful to a human, if sufficiently bribed, or can cause illness or even death. As the lightning deity's helpers they may be instrumental in recruiting shamans, causing various forms of precipitation, and guarding the earth lord's animals (the animals of the earth). There are both male and female rain dwarfs (Huber 1990), and a rain dwarf may become the "spirit spouse" of a shaman (Madsen 1955). In addition to human form, they may also take the form of a snake, lizard, or crocodile.

In the northern sierra of Puebla, Mexico, a female shaman named Juana was recruited to become a curer when struck by lightning while feeding her pigs. It frightened her and killed the pig next to her. It was rain dwarfs, locally called *tamatinime*, that had sent the lightning bolt. Later she related one of her experiences "with a *tamatini* (one who customarily knows things; plural, *tamatinime*). In this case, a *tamatini* appeared to Juana in the form of a female lightning bolt spirit. It was holding her male patient's soul captive in a cave, and Juana went to recover it" (Huber 1990:162). This too tells us that the rain dwarfs (or lightning deity's helpers) can be females as well as males. Juana was only one of the curers in Hueyapan, but all of them were recruited by *tamatinime*: "according to Hueyapan's curers, *tamatinime* forced them to assume their medical role after they appeared to them in dreams, or attacked them in snake, human, or lightning bolt form. *Tamatinime* appeared to most of Hueyapan's curers prior to or during adolescence" (Huber 1990:163). These rain dwarfs not only recruit shaman-curers; they are also believed to cause illness in other people. Called *chaneques* by Sierra Popolucas of Veracruz, they are described in ways that make it clear that they are assistants of the lightning deity, locally referred to as *Chane*.

> Nowadays humans are open to attacks from forces of wetness and coldness, and the best emissaries of such attacks continue to be chthonian spirits known as *chane*, *chanecos*, or *chaneques*, from *chan*, or *chanti*, a house. *Chaneques* dwell (*chantia*) in mountain tops, ruins, caves, water caverns (*atiopan* "water-god-site"), springs, and waterfalls. . . . Water *chaneques* (*achanes*, *taneko*) will sometimes take the form of ferocious alligators or water lizards (*acuetzpalin*), cousins of the forest lizard iguana (*baguetzpalin*). (Chevalier and Sanchez 2003:126)

Some Zapotecs in Oaxaca also see the lightning deity as having assistants. For example, in San Baltazar Loxicha Rosemary Beam de Azcona notes a belief in one main Lightning, with a number of little Lightnings serving him, perhaps thought of as apprentices or *topiles*. In one folktale from San Baltazar the little Lightnings are kept in covered jars, which are opened by a man who has been taken by a reptile to the realm of Lightning and other supernatural beings (personal communication 2011).

Thompson provides some evidence that the rain-lightning deities of Mesoamerica had a fondness for children, at least as sacrificial offerings (1970:171). He suggests that the Mesoamerican rain deities appreciated other things on a small scale, be they humans or utensils, providing evidence of the latter while connecting mushrooms and mushroom stones with the rain-lightning deities. Writing about the Aztecs, he says:

> In the sixteenth month, Coming of the Rains, the Aztec made small images of the mountains which were intimately associated with the Tlalocs, they laid very small tamales on tiny plates, they offered corn mush in tiny bowls and a very little cacao in small gourd vessels, and small boys were the singers at this feast.... In the recently discovered shrine of the rain gods of Balankanche, on the outskirts of Chichen Itza, large numbers of miniature metates, manos, and pottery vessels formed part of the offerings. (Thompson 1970:181)

As for the connection between smaller humans and the rain deity, we may recall the rain dwarfs that contemporary Nahuatl speakers see as servants of the rain god. Surely in bygone times earlier versions of these smaller versions of the rain deity were the venerated recipients of the tiny tamales on tiny plates, and the corn mush in tiny bowls.[17]

Early chroniclers in Mexico, including Sahagún and Durán, also pointed out the importance of children in ceremonies for the rain deity. Not only were children sacrificed to Tlaloc on Tlaloc's mountain (now named Mount Tlaloc); the children were encouraged to cry, by means that are unpleasant to contemplate, in the belief that their tears would attract Tlaloc and perhaps guarantee that he would supply the much-needed rain (Berdan and Anawalt 1997:154). Dwarfs and children share the element of size as well as a connection to rain. Nuyoo Mixtecs provide an implied explanation for the sacrifice of children to the rain deity. They consider plants, and particularly maize, to be "children" of the rain deities (Monaghan 1990:564), so it would make sense to sacrifice children of their own in order to gain maize "children" of the rain deities.

To summarize, dwarf beings, appearing to humans as little people, are found in the local lore of almost every region in Mesoamerica. They are called locally by different names and described in ways that make it clear they inhabit mountains, caves, and springs; interact with humans, giving rewards and punishments; and serve as assistants to the lightning deity. They can be male or female, and they recruit shamans, male or female. Human dwarfs were also depicted in Maya art and may have held a special place in Maya society because of their physical resemblance to various "rain dwarfs." Even today in some parts of Mesoamerica dwarfs are believed to be able to bring rain (**Figure** 41; see color plate section).

---

[17] If in some way dwarfs and children can be seen as miniatures, the fact that some indigenous altars can also be seen as miniature worlds is instructive, for as reproductions of significant aspects of the world it is noteworthy that they can embody the sacred mountain indexing the water and maize that it promises (cf. Sandstrom 2003).

The Lightning Deity's Animal Co-essences: Humans with Lightning Familiar

In Mesoamerica and elsewhere in the world individual humans are seen as tied by spiritual bonds to individual animals in the sense that they are born at the same time and what happens to the one during life will also happen to the other. So a person whose counterpart animal suffers a wound on its right hind leg while out wandering in the wilds will at the same time suffer a similar wound on the same leg. If the animal dies, the person to whom the animal counterpart belongs will also die. In some Mesoamerican societies under specifiable circumstances individuals can transform themselves into their animal "companions." In southeastern Mesoamerica the animal counterpart is often called a *nagual* in the local Spanish, and a more specific term is sometimes present in the local Mayan language. In Oaxaca the animal counterpart is frequently called a *tono* in the local Spanish and sometimes also in the local indigenous language. Both terms are borrowings from the Aztec language (Nahuatl), but in Nahuatl the *nahualli* refers to a transforming witch and the *tonalli* is a kind of soul linked in name to the heat of the sun, but also referencing the day-name on which one is born. A person's birth date, falling on one of twenty days, each having a specific name (sometimes that of an animal) and characteristics, is believed to influence one's own character, and so one's animal counterpart may be called a *tonal*, *tono*, or *tona* in contemporary Aztec societies as well. It is evident that the terms "animal familiar" or "animal ally" or even "animal spirit companion" could be seen as appropriate for this sort of relationship linking a human to an animal, even though "animal familiar" might carry connotations derived from Wiccan or other European earth religions, or perhaps other connotations not fully congruent with local meanings for the indigenous Mesoamerican terms. Although some of the above terms will be employed here, there is also a more neutral term for these animal counterparts that has been attributed to John Monaghan and employed by Houston and Stuart (1989) for Mesoamerican counterparts, as well as by Glass-Coffin (1998) in discussing spirituality and healing in northern Peru. The term is "co-essence," which is "an animal or celestial phenomenon (e.g., rain, lightning, wind) that is believed to share in the consciousness of the person who 'owns it'" (Houston and Stuart 1989:1–2; internal citation omitted).

The concept of co-essences or animal familiars is referenced in the literature on Mesoamerica with such labels as *tono, tonal, nagual, way, wayhel, lab, chon, co'k* (or *tzo'k*), *ombasneh, ser compañero*, "alter ego," "co-essence," "animal familiar," "animal spirit companion," "transforming animal," "animal ally," "guardian animal," and even "helper," "avatar," or "servant" (Freidel et al. 1993:196–197; Houston et al. 2001:450–457). Although one's animal familiar can be one of several or even many different animals (depending on one's specific community of origin), there appears to be peripheral evidence of a relationship between snakes and the generic notion of an animal familiar or co-essence. This may derive from the associated symbolic relationship of transformation, a quality often attributed to serpents because they periodically shed their skins. Tzotzil *chon* "animal familiar" also means "snake, worm"; Tzeltal *lab* "animal familiar" is likely a cognate of Jacaltec *lab* "omen," which is itself likely etymologically related to Jacaltec *laba* "*culebra, salamandra/snake, salamander*," which can be compared with Mam *labaj* "snake." Huastec Maya has a word *labax* "*fantasma, duende, agüero*/phantasm, dwarf, omen." Chuj, another Mayan language, has *lab* "*nagual*, transforming witch, bad omen." This evidence of a generic relationship between snakes and animal familiars could be related to an Aztec notion connecting all of the three human souls to serpents as illustrated in an Aztec codex: "In Codex Laud, the souls are represented as three serpents emerging from the body. The yolia erupts from the mouth, the tonalli from the top of the head, and the ihiyotl from the intestines. Here, they leave the body at death, and each goes its own way to a different fate" (Furst 1998:213). The generally shared notion of animal familiars in Mesoamerican worldviews is that every human being has an invisible connection to a particular animal by means of a spirit or essence that they share (Pitt Rivers 1970; Furst 1998; Houston et al. 2001). The fortunes in life of the human are simultaneously those of his or her animal familiar (co-essence, animal spirit companion) (Ruiz de Alarcón 1982:64; Vogt 1965). A corollary to this notion of animal familiars is that certain individuals are able to literally transform themselves into an animal, and in many cases this is believed to be their own companion animal. For example, among sixteenth-century Aztecs in central Mexico,

> if the person was capable of changing into animal or meteorological shape, he or she was a *nahualli*, and the Mexica also applied the term to the animal or meteorological alter ego. On the other hand, if the thing came into the world at the same time as the human being, suffered the same fate, and died at the same time and in the same way as the man or woman, while keeping its own form, then it was one's *tonalli*. In that instance, person and animal shared a common spiritual essence and fate, but not the same body. Instead, their lives followed parallel courses. (Furst 1998:216)

Communities differ as to whether or not everyone has an animal familiar, whether one can know which animal is one's companion animal, how to find out one's own animal familiar's identity, whether transforming witches are important or even present, what mischief (or good) can be accomplished by one's familiar, and so on. For example:

> In northern Puebla, many people agree that everyone has a *tonalme*, an animal companion spirit, but that some, the *nagualme*, can literally change forms and become animals. Members of the latter group alter their forms in order to punish enemies or to chasten those who behave

> badly. . . . The people of Huitzilan de Serdán, Puebla, believe that every human being has a *tonalli*, or animal, that "corresponds" to his inner nature. The identity of this creature is generally unknown to the individual, but should it be injured, become ill, or even die, the same will happen to its human co-essence. On the other hand, *hechiceros*, or "sorcerers," physically transform into their animal alter egos. (Furst 1998:216)

The Nahuatl- derived terms *nagual* and *tono* (or *tonal*) are commonly used throughout Mesoamerica, having diffused shortly before or shortly after the Spanish Invasion, and Foster (1944), because of this commonality, recommended distinguishing transforming sorcerers from individuals connected to animals through a common soul by using the terms *nagual* and *tonal*, respectively, for these concepts. There is enough variation in usage from community to community, however, to render these terms problematic when generally applied according to Foster's suggestion (Paz 1995). For example, in the Quiché Mayan village of Santiago El Palmar people variously use the term *nagual* for signs of the zodiac, days of the 260-day sacred almanac, the patron saint of the village, and/or the spiritual essence of the earth (Saler 1964:310–311). In the Tzeltal Mayan community of Tenejapa, *nagual* and the native term *lab* are applied to a person's animal familiar (be it animal or meteorological), to a transforming witch, and also to a class of phantasms that scare people on the trail at night, such as a skeleton known as Flesh Dropper, or a bodyless head known by some as Charcoal Cruncher that rolls along the paths. These latter are co-essences or alter egos of human individuals who appear to be normal people by day, but that under certain circumstances, such as at night, they transform into the other form with which they share an essence or soul. Such variability regarding the meanings of *nagual* suggests that, as general terms for making the distinction proposed by Foster, *nagual* and *tonal* are less useful than one might expect.

An additional corollary of the notion of animal familiars so widespread in Mesoamerica is that the familiars linked through a shared soul or co-essence are not always animals. In some cases they are meteorological phenomena that might appear to be alive, such as lightning, whirlwinds, or "comets" (Miller 1956: Lipp 1991; Furst 1998). In these instances certain individuals can transform themselves into the meteorological phenomenon with which they are linked. Needless to say, they are considered to be particularly powerful individuals, and in many cases they are thought of as witches or sorcerers.

If humans can have animal familiars, it stands to reason that anthropomorphic, and named, entities controlling natural forces, here called gods or deities, might also have their own co-essences or animal familiars, which indeed they do throughout most of Mesoamerica (Spero 1987, 1991; Houston and Stuart 1989; Lipp 1991; Monaghan 1995). Some of our best evidence for this comes from post-Conquest oral narratives and from images of pre-Conquest rain deities.

God B of the Postclassic Maya codices is recognized as the contemporary rain-lightning deity Chaak, and believed also to be the descendant manifestation of the Classic period Maya rain-lightning deity Chaak. In the codices he appears to have some snake features, including a long and often pendulous nose or upper lip, and he brandishes an axe or a serpent. One might view the former as symbolic of a thunderbolt and the latter as symbolic of a lightning bolt. In many Classic images Chaak has serpents issuing from his mouth, and sometimes he wields an axe.

God K of the Maya codices (corresponding to Classic Maya K'awil), although distinct from codical God B (Classic period Chaak), shares some features with him, including a long snout or nose, associations with serpents, and associations with an axe. Their Classic period counterparts K'awil and Chaak are not so easy to distinguish from one another, and even if distinct, they may well represent different aspects of the same lightning-rain deity according to Coggins (1975, 1979, 1988) and Taube (1992a, 1992b). Regardless of the ambiguities with respect to codical gods B and K, and also to Classic period Gods Chaak and K'awil, the Mayan lightning deity of the Classic period certainly had a serpent as his animal familiar.

Just as certainly, at least one of the contemporary lightning deity's animal familiars is a serpent. According to Chamula Tzotzil Mayan tradition, caves are inhabited by the earth lords (the lightning/rain deities) and by snakes, "which are familiars and alternative forms of the Earthlord.... Hence, they are associated with dampness, darkness, and lowness" (Gossen 1974:21). But it is not unanimous that serpents are the animal familiars of the lightning deity. The lightning bolt itself may be his "familiar." A few miles from Chamula, in Chiapas, Mexico, is another Tzotzil-speaking community, San Pedro Chenalhó, where it is believed that the earth lord/lightning deity, known locally as Anhel, "is the owner of the thunderbolt produced by pounding on his drum. He appears in dreams as a saint image or a man. Some nights he is perceived as a head rimmed by fire and with a long tail. His instrument is the lightning bolt, which is said to be his *wayhel* or co-essence" (Brady and Prufer 2005:169–170). The root *way* of this Tzotzil term for the co-essence or animal familiar is the same as the Classic Mayan glyphically spelled word for co-essence, *way*, found painted on pottery as well as other surfaces (Houston and Stuart 1989).[18] This attests to the antiquity of the concept in Mesoamerica, its relation to the concept of sleep (in that *way* also means "to sleep"), and also to the spatial and temporal continuity of the word itself, since Tzotzil Mayan peoples inhabit an

[18] The *hel* in this word may well be Tzotzil *hel* "change, exchange, substitute" (see Laughlin 1975:150).

area considerably removed in both space and time from the region occupied by the lowland Classic Maya.

Simultaneous in birth and in death, it is because the lives and fortunes of a human and his animal familiar are so closely connected due to a shared soul or essence that the concept has considerable meaning and force for many indigenous Mesoamericans. This was true in centuries past, and it still is today. Most individuals today are thought to have a single animal familiar, but it seems that at least among the Maya in Classic times an important person, such as a ruler, might accumulate multiple animal familiars, adding to his own those from enemies that he has killed (Houston et al. 2001; Houston and Stuart 1989). That individuals in some communities today can possess multiple co-essences is a tradition attested in the Tzeltal Mayan community of Pinola, where every individual is supposed to have at least three different animal spirit companions, up to a maximum of thirteen (Hermitte 1970:80).[19] It is also attested in the Mixtec community of Nuyoo:

> Nuyootecos say all humans maintain a nuvi-like connection between at least two distinct states of being—a human body on the one hand, and an animal body, one's *kitï nuvi*, or "'transforming animal," on the other...when a human is born, an animal is born, out in the forest. This creates an intimate bond between the two, since they share the same ta'vi, or destiny. Thus when one feels hunger, they both feel hunger. When one hurts, they both suffer the pain. The connection extends even to death, so that when one of the two dies, the other also dies. Some people have several *kitï nuvi*, each of different species. Such persons stand a better chance of a long life than someone who has only one *kitï nuvi*, because the person with several animals only sickens if one of them dies. (Monaghan 1995:347–348)

Currently in a number of Mesoamerican societies, including those of various Mayan groups, Mixes, Huaves, and Zapotecs, there are ways of finding out what kind of animal one's own familiar is. Often an individual will discover the animal's identity in a dream, but there are other ways too. For example, when a Tenejapa Tzeltal woman is about to give birth, her husband or another relative will sweep carefully around the house in which parturition is taking place. Once the infant is born the swept ground will be examined for tracks, scat, or other spoor of an animal, and whatever animal is identified as having left its mark will be the animal spirit companion or co-essence of the infant. This custom, now disappearing, is the basis for a Tzeltal Mayan disparaging joke still circulating about Chol speakers living to the north of them, one of whom

[19] The number thirteen, corresponding to the thirteen layers of the sky in Yucatec and some other Mayan traditions, also corresponds to the number of parts that the human soul, or essence (*ch'ulel*) is believed to have in the Tzotzil Mayan community of Zinacantán.

supposedly swept all around the house where his wife was giving birth. The district postman delivered the mail on his bicycle during the birth, so shortly thereafter the Chol man announced to the family that his newborn son's "animal familiar" was a bicycle.[20]

Based on the *Descripción geográfica* of Francisco Burgoa, a Mexican historian of the Dominican order born around 1600 in Oaxaca, Bancroft (1900:661) notes a Zapotec tradition in which relatives of a woman about to give birth would assemble in her domicile and start drawing simple figures of animals on the dirt floor. As each figure was completed it would be rubbed out and the figure of another animal begun. This procedure continued until the woman gave birth. The animal whose figure was being drawn at the moment of birth was then ascertained to be the *tono* of the newborn infant. According to that Zapotec tradition, "when the child grew old enough, he procured the animal that represented him and took care of it, as it was believed that health and existence were bound up with that of the animals, in fact that the death of both would occur simultaneously" (Bancroft 1876/vol. 1:661; see also Frazer 1900/vol. 3:411).

One way that a Mixe family in Oaxaca discovers the identity of a newborn's *co'k* (*tono*, guardian spirit, or animal familiar) is to intertwine filaments of an agave plant and put them in a cup of clean water on a table in the middle of the house where it is spoken over a day before the birth. After the birth the filaments will be examined for footprints of the infant's *co'k* (Lipp 1991:118). More commonly the infant's father puts a small portion of lime or ashes in each of the four corners of the house. Upon birth a track of the infant's familiar will be left in the lime, and in some cases the track will be spiral-shaped whirls on all four heaps of lime. This indicates that the infant's *tono* is in fact 'Ene·, the lightning deity.

A Huave speaker, desiring to ascertain the animal identity of their child's *ombasneh* (*tono*, co-essence) will place a shallow bowl of sand or ashes, with the surface smoothed, covered with a cloth, under the bed of the mother-to-be

---

[20] A Mexican film distributed in 1954 and named *Raíces* devotes an episode to the Chamula Tzotzil practice of putting down lime powder or ashes (rather than sweeping) in order to view the tracks of a newborn's animal spirit companion for the purpose of naming the infant. The film attributes the practice to Tzotzil speakers rather than Chols, who are said to lack the practice, and in the film a bicycle is mistaken for the infant's animal spirit companion. This researcher heard a variant of the bicycle story as a joke in 1968 told by a Tzeltal speaker about Chols, and a 2011 query to a Tzeltal consultant yielded the information that some tell the joke about Tzotzil speakers. It would not be surprising if Tzotzils also tell it about Tzeltal speakers. While at least one anthropologist believes that the film is where the story about the bicycle originated, it seems more plausible for a film to portray an indigenous practice than for an indigenous practice to be borrowed from a film. Few if any Chols, Tzeltals, or Tzotzils saw movies during the 1950s in any case; but to be safe one may reserve judgment. In any event, the practice of trying to identify a newborn's animal spirit companion is found in a number of Mayan and non-Mayan societies in Mesoamerica, and sometimes the co-essence is found to be not an animal but rather one of the lightning deity's helpers. It is probably never really believed to be a bicycle.

near the time of birth. After birth the bowl is uncovered and the sand or ashes medium is examined for the spoor of the newborn's animal counterpart. Some Huaves say that this could be a Mixe custom.

Stories abound, of individuals whose familiars are fighting each other or are suffering some misfortune in the wilds (Ruiz de Alarcón 1982; Corlett 1934:196; Vogt 1969), or whose familiars revenge an insult or injury by killing the culpable individual's livestock (Turner 1972:59). In many contemporary communities an important person, and especially a shaman, is supposed to have a jaguar as his familiar, and there are communities that believe the shaman can transform himself into a jaguar, inhabiting the body of his own familiar (Vogt and Stuart 2005:176), and of course thus possessing the great power of that predator. "The jaguar is the most powerful animal companion and always serves as the spirit companion of a Chol Shaman" (Vogt and Stuart 2005:176).

The capacity for transformation into one's own animal familiar is illustrated in the Colonial period Quiché account of a battle with Pedro de Alvarado during the invasion of Guatemala, when a Quiché captain whose familiar was an eagle lost a battle to Pedro de Alvarado, who was defended by *his* familiar, a "very fair maiden" (Bricker 1981:39). It is of interest that the Colonial period document remembering the event did not say that either of these leaders had a jaguar animal familiar; however the fact that shamans are often reported to have jaguar familiars, brings the jaguar (as well as the eagle) into association with lightning as asserted below.

In many communities today some individuals are thought to have "animal" familiars that are in fact not animals but active natural phenomena, such as a thunderbolt, whirlwind, ball lightning, or meteor.[21] These individuals are particularly powerful. It is certainly relevant here that "war leaders of numerous Maya groups are reported to transform themselves into lightning in order to attack enemies. For example a K'iche' captain named Izquin Ahpalotz Utzakibalhá, or Nehaib, assaulted the conquistador Pedro de Alvarado in this form" (Looper 2003:84; internal citation omitted: see Bricker 1981:40). Miller says that Mixes in Oaxaca retain a belief that a shaman can take the form of a lightning bolt. Though it may look like a natural lightning bolt, this kind is known as a *nagual rayo* to distinguish it from a natural lightning bolt. The Mixe animal familiar, known locally as a *tono* or *nagual*, is seen as a guardian spirit or companion being that is intimately tied to the life and fortune of the

---

[21] Individuals in some communities have familiars that cause fear in people: such individuals as Tzeltal Flesh Dropper who goes out at night and strips off his flesh at a trailside cross to become a skeleton walking the trails and scaring people who shouldn't be out at night; or the individual whose head leaves her body at night and wanders around scaring people on the trails; or a female bloodsucking vampire; or the Huastec converting into a transvestite phantasm that wanders the trails at night scaring people, or the *paslam* fireball (person) that refers to ball lightning, meteorites entering the atmosphere, or meteors (see Taube 2000:325).

person. When a person transforms into that which is also his *tono*, a *rayo* (lightning bolt) for example, it is said that the person has a *rayo* as his *tono* or his *nagual* (Miller 1956:138), and the person when transformed is then also referred to as a *nagual* (Miller 1956:225).

In the southern mountains of Oaxaca, San Baltazar Lochixa Zapotecs accept the idea that people are born with a *tono*, which refers to the animal familiar or co-essence that is elsewhere called a *nagual*.[22] While most tonos are animals, specific people may have a lightning (*rayo*) tono. "It is believed that you can pray to your tono and that your tono will avenge you. So when someone is struck by lightning people will think back to who recently had a fight with the person, and deduce that this other person has a '*tono de Rayo*'" (Rosemary Beam de Azcona personal communication 2011).

The assistants of a rain deity may also have spirit familiars. The following passage is evidence of perceptions by Nahuatl speakers in Puebla's northern Sierra regarding the rain-lightning deity's assistants.

> Hueyapan's curers claim that *tamatinime* played an important role during the liminal phase of their recruitment. *Tamatinime* are thought to be wise and powerful spirits who live in caves, and frequent streams, waterfalls, forests, mountainous areas, and the ocean. They are referred to by several additional names depending upon the form they take and the function they perform As rayos (lightning bolts), they punish people with illness who show disrespect for nature or for other people. As *achihualime* (rainmakers), they appear as small naked children with light curly hair. San Miguel Arcangel and the Virgen del Rayo (also known as the Virgen del Carmen) command seven male and thirteen female rainmakers, respectively. Tamatinime may also appear as snakes, men, and women, and are sometimes thought to be actual individuals with lightning bolt companion spirits. (Huber 1990:163; internal citations omitted)

To sum up, deities, their assistants, and humans can all have familiars, co-essences that share each other's fortunes with them, and into which they can sometimes transform themselves. The lightning-rain deity—whose name in some communities means "lightning" and in others means "thunder," "owner," "lord," or "grandfather"—has been associated with two different familiars, serpents, or lightning bolts, according to different communities. In the Isthmus of Tehuantepec and the southern mountains of Oaxaca, Mexico, it seems that iguanas or crocodiles are also possible familiars for the lightning deity. His assistants may appear as, or be transformed into, snakes, lizards,

---

[22] Both *tono* and *nagual* are indigenous usages of Spanish borrowings from Aztec (Nahuatl) *tōnalli* and *nāhualli*, which refer in Classical Nahuatl to "day, heat of the sun" and "sorcerer" respectively (Karttunen 1983).

lightning bolts, human children, or human adults of either gender. These assistants may have their own assistants, which in turn may appear as (or transform themselves into) snakes, lightning bolts, children, or adults of either gender.

Humans have a variety of different animals or a small number of natural phenomena as possible familiars. Most people cannot ordinarily transform themselves into their animal familiar, but in Mesoamerica especially powerful people, such as important warriors and shamans, can effect such transformations when desired, and sometimes do so involuntarily. The most powerful people have a jaguar or a lightning bolt as their familiar. In some communities the shaman is associated with a jaguar familiar; in others the shaman's familiar is a bolt of lightning (a thunderbolt). The notion of deities related to natural phenomena that exist in multiples and that have assistants, and the related concept of familiars and of transformation into one's familiar, are part of what has been called the Mesoamerican shamanic tradition or shamanic cosmology.[23]

Lightning, Warfare, Protection, and Punishment

We have seen above that certain persons are believed to be able to transform themselves into lightning bolts in battle—if the lightning is their co-essence—in order to overcome the enemy with its power and potential for great destruction. Thunder accompanying lightning also inspires fear. Inspiring fear and punishing an enemy is at the same time both punishment and protection for it is an activity concerning the protection of oneself and friends.

The power to inspire fear, exact revenge, destroy enemies, and protect friends is of course something very useful to a king or other leader, so it is understandable that a Mayan ruler might seek to be portrayed on public monuments as related to lightning—that is, as both destroyer and protector.

> In assuming the ritual identity of a lightning god, K'ak' Tiliw [a Quirigua ruler] drew upon deeply rooted associations of lightning with warfare. Indeed one of the most common war titles in the Classic period was *kalomte'*, depicted as a Chaak, holding the sacrificial axe. Related beliefs survived the Spanish invasion. One of the most venerated Jacaltec culture heroes is El K'anil, the "man of lightning," who destroys enemies by transforming into lightning. Among the Ch'ol the thunderbolt spirit Chahk protects towns from evil. The Tzotzil tell many tales about Chauk or Thunderbolt, who destroys enemy war-

---

[23] Frazer (1900:407–423) presents numerous examples from many different parts of the world, from Australia to West Africa, to Siberia, and the Melanesian New Hebrides, as well as Australia, in which individuals are tied to animals through shared souls, and the death of one spells death at the same time for the other. Just as in Mesoamerica, there are numerous variations on this general theme.

> riors with an electrical storm (Looper 2003:84; internal citations omitted)

Kings no longer rule in Mesoamerica, but shamans remain, and they fulfill both symbolically and in actuality a number of the functions that kings used to do symbolically. Two of these functions are represented by the delivery of reward and punishment. Spero provides considerable evidence that in the Mayan region "lightning is the protector of the community and the crops" (1991:189, see also Aulie and Aulie 1978:46), further explaining that Mixe-Zoquean as well as Mayan individuals having lightning familiars also retain the power to direct lightning toward individual punishment in addition to using it for community protection (1987; 1991:189). Bassie-Sweet (2008:7) attributes to David Stuart the suggestion that in Classic times Maya sorcerers, through their familiars, were believed to send disease and other misfortune, presumably including such misfortunes as could be attributed to lightning and lightning familiars.

In the central Mexican highlands, Classic period Teotihuacan had a lightning-rain deity with "goggle eyes" that was as closely related to warfare as it was to rain and that was represented in transformation by a "war serpent" (Taube 1992b; Freidel et al. 1993:296). The Postclassic Aztecs, later arrivals to the same region, appropriated this deity (with similar appearance and functions), named him Tlaloc, and reduced his connections to war, placing greater emphasis on his control of rain, including the destructive capabilities of lightning, hail, deluges, and other weather phenomena. The Classic Maya, however, borrowed iconography pertaining to this deity from Teotihuacan, including images of the central Mexican spear thrower, to serve primarily in the representation of warfare.

Nuttall (1891) refers to the finger holes of the spear thrower, a war instrument known in Nahuatl as an *atlatl*, as the "eyes of Tlaloc" that they are said to symbolize, Tlaloc being a convenient appelation for the Classic period central Mexican lightning-rain deity, who was also a war deity borrowed by the neighboring Classic Maya to portray warrior rulers. She notes further that in the sculptured monuments "we find the atlatl and the spear and an almost total absence of any other weapon" (Nuttall 1891:29). So the lightning deity, his serpent familiar, and the weapon symbolizing him in iconic form, are all integrally related to war, both in the central Mexican highlands and in the Maya region during the Classic period (Taube 1992b). The Nuttall Codex, a Mixtec screenfold picture book made in the fourteenth century but depicting events of a century or more earlier, also illustrates well the association between the rain-lightning deity and war (Williams 2009:108).

Taube also maintains that for the Maya "both the Classic and Post-Classic Chaks are closely identified with war" (1992a), the Chaks being the native Maya representation of the lightning-rain deity also known as Chaak. Amplifying this identification, he looks to the war serpent. "In Classic Maya iconography, the War Serpent is consistently identified with fire. Thus it has been noted that flames frequently exude from the mouth of the creature. A Late Classic full figure glyph from Copan depicts the War Serpent as the serpent foot of God K. In this case, God K is rendered as its Mexican counterpart, Tlaloc, another god of rain and lightning. The War Serpent in turn replaces the conventional Bearded Dragon serving as the flaming foot of God K" (Taube 1992b:64–65). Spero, looking at vocabulary rather than iconography, points out that "among the Mayan languages having linguistic associations between Kawak and weapons, we find that in Yucatec the term for 'flint' is *bat chaak,* the axe of the Rain God *chaak*" (1991:188; internal citation omitted).

Linkage of the Classic period Zapotec lightning deity, and his Colonial period variant, Cocijo, to warfare is difficult to verify, primarily due to a lack of extant records. Among today's documented Zapotecs, Cocijo is most commonly referenced with respect to lightning, rain, hail, frost, and wealth. Though in contemporary folklore he does punish as well as reward, persuasive evidence of his relationship to warfare in the pre-Columbian iconography is scant.

Lightning is not only a destructive force for the enemy on the battlefield. Ordinary farmers also see it today as potential punishment for insufficient attention to sacrifice and prayer to the Earth Owner. Villa Rojas has discussed a devotion to caves on the part of Oxchuc Tzeltal Mayans that "has its origin in the belief that it is from them that lightning comes to punish and to disperse the natural elements (hail, winds, storms, etc.) that frequently threaten and even destroy the corn fields" (1947:579). Monaghan (1995:109–110) discusses a Mixtec community in Oaxaca where lightning is used to punish a transgression like vandalizing someone's cornfield. Numerous ethnographic works on Mesoamerican societies provide ample evidence that the lightning deity punishes offenders, sometimes with lightning storms, at other times with not enough rain for their crops.

Returning to the battlefield for a moment, apparently over the centuries many military leaders unrelated to Mesoamerica have recognized that rains seem inevitably to follow a heavy battle, as evidenced in the following passage: "There is a very prevalent belief that great battles and heavy discharges of artillery are followed by rain-storms. I recently read what purported to be a scientific article on the causing of rainfall by mechanical disturbance of the atmosphere. The author delved into history to prove his theory. He showed that all the great battles of the Civil War as well as of other wars were followed by rain-storms" (Guinn 1903–1905:175).[24] Regardless of the accuracy of this

---

[24] This is further indicated by an excerpt from a letter written during the civil war by Union General Carl Schurz: "A heavy rain set in during the day—the usual rain after a battle and large numbers had to remain unprotected in the open" (Billington 1962:396). Ambrose Bierce wrote, of the battle of Shiloh, "The night was now black-dark; as is usual after a battle, it had begun to rain" (Bierce 2007:120).

belief, its indexicality and iconicity with respect to the relationship between thunder and lightning and rain recalls the ceremonial ubiquity of present-day skyrockets and other fireworks in Mesoamerica, and the frenetic whistles, shouts, sounding of conch shell trumpets, and booming of drums that accompanied battle for indigenous Mesoamerican armies as noted by the Spaniards during and after the Invasion (Diaz del Castillo 1981:21, 32, 47, 73, 144, 325, 330, 343, 381, 404). So in this way there appears to be a relationship between the similarity of thunder and lightning to the thunderous noises of combat in war that the Mesoamericans may well have related to rain and the rain-lightning deity, bringing war and rain together in a single deity. This is suggested in the following passage about a contemporary Tzotzil Mayan community: "In Zinacanteco belief the Earth Owner rides on a deer and uses iguanas as blinders for his mount and a snake for a whip. The shell of a land snail is his powder horn for making skyrockets and shooting his shotgun, both of which are the bolts of lightning" (Vogt and Stuart 2005:164). In summary, lightning has clear associations in Mesoamerica with warfare, one aspect of which is destruction of the enemy. The other side is protection of oneself and friends or subjects. Analogously, the lightning-rain deity administered both punishment and protection through destruction or nourishment of crops. Nourishment of crops brings us to the next heading.

Lightning Deity, Crop Fertility, and Wealth

While warfare serves the two functions of vanquishing the enemy and protecting the friend, protection itself leans in the direction of providing sustenance for the friend (or one's subjects). The king, ruler, shaman, or other leader in Mesoamerica is, and was, charged with all three of these duties. Indeed, rulers in many countries, in addition to making war for conquest or self-defense, have been expected to sow the first grains or roots of the staple crop in agricultural societies (witness ancient Egypt, West Africa of a century ago, or modern Japan). Close connections between the king and the rain-lightning deity are therefore to be expected. In Mayan lore, for example, "it is thought that leaders are capable of using thunderbolts to defend their communities and that lightning strikes germinate the corn seed" (Bassie-Sweet 2008:6–7).

With proper veneration through respect, ritual and sacrifice the lightning deity rewards the farmer by creating conditions for rich harvests of desired foods. Maize is specifically implicated here, but of course beans, squash, various greens, and other Mesoamerican food crops are included as well. Such good conditions have to do primarily with rain at the right time and in the right amounts, but other things are also required for harvesting the desired foods, whether these foods are derived from plants or from animals. With agricultural products the seeds need to germinate, and the weeds should not be too thick. The soil must be fertile, preferably not too rocky. The lightning deity is also implicated in provisioning from animal foods. Many factors ensure success in hunting, for example, and throughout large areas of the Americas there is an entity facilitating or forbidding success in the hunt, often called the "Master of Animals." In Mesoamerica the Earth Owner rain-lightning deity is usually this same "Master of Animals" (cf., e.g., Foster 1945:186). Furthermore, this deity is also associated with treasure that is neither vegetal nor animal. He is said to preside over various forms of treasure, including gold and money, in his abode (Lipp 1991:48), and he is not infrequently linked to treasure tales, including those generated to explain how someone who is believed to have gotten suddenly and recently rich got their riches.

Lightning is the only native deity still recognized through ritual—as opposed to simply in mythology—by the Zapotecs of Mitla, Oaxaca. He is the god of rain, fertility, and abundant crops. He can appear in the form of a human, or more frequently in the form of a lizard (Parsons 1936:212–213), and some say an iguana. He punishes transgressions with hail and rewards good behavior with rain and good weather. In the southern mountain region of Oaxaca in which Zapotecan languages are spoken, the lightning deity is also associated with reptilians such as the crocodile or the serpent, but most typically he is associated with the iguana (Rosemary Beam de Azcona personal communication 2011).

The fourth day-name in the Zapotec 260-day calendar is given as either lizard or frog. Both of these are connected with rain and the lightning deity in Mesoamerica. In the Aztec twenty-day calendar the fourth day is lizard or iguana, both of which symbolize wealth (which to a farmer would mean a good crop with a surplus). In the Maya twenty-day calendar the fourth day is K'an, a word that means "yellow" and "ripe," and the glyph for this day is said to be a kernel of maize (Taube 1989).[25] The fourth day-name then constitutes one more way that lightning is associated with wealth, not only in that lightning is connected to maize and lizards in the 260-day Mesoamerican almanac, but also in the sound of the name for the fourth day, which sounds similar to and is cognate with the Jacaltec name for the eighth *veintena* day, Q'anil "yellow," symbolizing maize and money, and also the name of Lightning Man in Jacaltec oral tradition. He is protector of the community.

That serpents may also be associated by Zapotecs with lightning can be inferred from such data as the fact that near Mitla, Oaxaca "there are carved stones which are held to represent serpents which came to life and defended the town against the invading Spaniards" (Parsons 1936:224); or, from another narrative, "at Matatlán I was told that a certain rich man at Tlacolula is rich because of the snake he keeps in his trunk" (Parsons 1936:224). Maize is

[25] It may be a red herring, but the Mixe neighbors of the Mitla Zapotecs have a name for a kind of salamander, the *tlaconete*, which is ⟨añu cuda'ax⟩, and while the first word means "thunder," the second is very similar to the word for "owner" or "lord." It lives in the cloud forest.

included in the relationship in that "corn maidens figure in the tale of Lightning and his *compadre*" (Parsons 1936:228), and "sacrificial bread is made from the flour of double ears of corn (*kwachi*, twin; or *niskwach*, corn ear, twin)" (Parsons 1936:228). The term *kwachi* is originally from Nahuatl, where it refers to both snakes and twins.

In highland Chiapas, Mexico, a shaman consultant told Calixta Guiteras-Holmes about the earth lord/lightning god known as Anhel in the Tzotzil Mayan community she studied, saying, "'Anhel is the rain god, the lord of the mountains, the corn giver, the master of beasts, and the god of the waters. He watches over the cornfields planted around the foot of the mountains and on the summits and procures the necessary rain for them'" (Guiteras-Holmes 1961:60). In Mayan mythology it is lightning strikes that germinate the corn seed (Christenson 2001:74; Bassie-Sweet 2008:7).

The connection in Mesoamerica between lightning and fertility, and abundance of maize, is a strong one and an obvious one, but there is a dimension to it that is not so obvious even though it derives from careful observation. Lightning is the only natural nonbiological means of fixing nitrogen to make it available for plant fertilizer. Many have noticed that after a thunderstorm, and particularly near where lightning has struck, plant life, especially fast-growing grass (of which maize is a type) seems to get a growth spurt. Plants need to secure their nitrogen, a requirement for life, in "fixed" form (i.e., made available by being incorporated in such compounds as nitrates, ammonia, and urea), which is mostly done with the help of microorganisms. Lightning has the energy to break the very strong bonds between nitrogen atoms in the air, producing molecules that are usable by plants even without the help of microorganisms. Depending on geography, lightning may account for some eight percent of the total amount of nitrogen fixed by natural means and thus is of some importance in plant growth. Apparently this was recognized by Mesoamericans, just as it is today by Western science.

In short, lightning itself is directly connected to maize growth in that it produces a vital fertilizing component by making nitrogen available in usable form to plants. Another association is the perceived relationship between crop fertility and the lightning deity; an important connection in that it is he and his helpers that reward farmers with successful *milpas* (cornfields) or punish them in various ways. Wealth and lightning are linked in treasure tales involving the Earth Owner (lightning deity). The greatest treasure of all in indigenous Mesoamerica is maize, the key sustaining staple food according to native perceptions.

Lightning Splits Sustenance Mountain, Bringing Maize to People

Yet another connection between lightning and fertility, abundance, and specifically maize is made in a significant portion of Mesoamerica. This connection is apparent in the widespread Mesoamerican narrative (often referred to as a "myth") concerning lightning splitting open Sustenance Mountain so that humans could get at the maize within and thus plant and consume it.

The most basic aspect of this narrative is that lightning splits open a rock, mountain, or turtle shell that contains maize within it, making it available for humankind. The Classic Maya called this rocky mountain Yax-Hal-Witz "first-true-mountain," and the Colonial period Quiché Mayans called it Paxil "Split Place" (Freidel et al. 1993:111; Bassie-Sweet 2008:7–8). Hundreds of miles to the northwest, a related Huastec Mayan narrative tells of a mountain named Paxal, split open by a lightning bolt of the lightning deity (Muxi' Maam), revealing for people the maize that was stored inside (Alcorn 1984:62). Tzeltal Mayans tell of a stone mountain being entered long ago by a blast of lightning from a bird named Ala Witz Mut (after a woodpecker had first knocked at the door of the stone mountain),[26] revealing the maize stored there (Stross 1973:104; Slocum 1965), and a Chol myth "reported by Manca (1995:225–226)...describes how Ch'ujtiat first found maize in a cave. A bolt of green lightning cracked open a rock under which he discovered kernels of maize" (Vogt and Stuart 2005:175–176). Speakers of Mopan in San Antonio, Belize, have a long tale in which can be found many motifs found also in the sacred book of the highland Guatemalan Quichés, the Popol Vuh. In the Mopan narrative a woodpecker is sent by the chief rain deity to tap on a huge stone under which maize was hidden in order to ascertain where the stone was thinnest. Then Yaluk threw his bolt of lightning at the thinnest part and broke open the rock, revealing maize, which was eventually obtained by humans (Thompson 1930).

In some versions of a central highland Aztec narrative of the discovery of maize Quetzalcoatl transforms himself into an ant in order to discover maize and other food plants "where they lie in the heart of a mountain, which the lightning god then splits open so that the rain gods can extract them" (Krickeberg 1968:42). Pipils migrating after the breakup of the Toltec Empire made their way to El Salvador, speaking a language closely related to Classical Aztec. They tell of "rain boys" who live on mountaintops, pour water onto the earth as rain and "like the Aztec lightning god, they once split open a mountain in order to extract the seeds of the maize plant" (Krickeberg 168:82). Contemporary Nahuatl speakers from the southern Huasteca say that Woodpecker was called upon to open the Mountain of Sustenance,

---

[26] It is notable, in view of the woodpecker's attempt to get at the stored maize, that a San Juan Ixcatlán Mixe name for the imperial ivorybill woodpecker that announces rain with his pecking is "thunder's chicken" (Lipp 1991:29), while a Coatlán Mixe term for another woodpecker species is "thunder's paint-striped one" (Hoogshagen and Halloran 1993:58). It is also of interest that I was told by a Tzeltal storyteller that Ala Witz Mut is an eagle, suggesting a relationship between the eagle and the lightning deity. "That 'Ene· can assume the form of an eagle or serpent clearly relates this deity to the Tlaloc on the Tepantitla mural" (Lipp 1991:29) even more directly relates the Mixe eagle to the lightning deity.

called Postectli or Postectitla (Broken Hill), so people could get at the maize within (Sandstrom 1991:241, 246, 250). After Woodpecker had tried and failed, old San Juan split the mountain into seven pieces with a lightning bolt, revealing and partially burning the maize (Read and Gonzalez 2002:177–178).

Although early Zapotecs may have maintained an explanatory narrative concerning a Sustenance Mountain and the discovery of maize for human use (Joyce 2009:134, 174), the remaining evidence is scant, and even so there is little to implicate lightning in splitting open that mountain. There are a number of Zapotec lightning stories from Mitla recorded, and one of them involves a man who is Lightning's compadre going up to Lightning's cave on the mountain to get some maize for himself and his brother (Parsons 1936:329) but there is nothing about splitting open the mountain. A Zapotec explanation from San Baltazar Loxichá may reflect a modern belief with earlier antecedents implying lightning strikes by different lightning beings with different strengths prior to a successful conclusion, as in the lightning splitting open of Sustenance Mountain. Rosemary Beam de Azcona was told that lightning had hit an electrical box nearby and that the box had kept humming for several days until Lightning struck it again, and then the noise stopped. "The interpretation was that it had first been struck by Little (perhaps adolescent) Lightning, who then became trapped in the electrical box. The humming was him trying to get out. Then 'his friend' came and got him, when the more powerful Lightning struck days later and released the inferior Lightning from the box" (personal communication 2011).[27]

A Oaxaca Mixe story collected by Miller (1956:110) tells of a young woman who went deep into a cave on the mountain where she met and married Rayo (the thunder-rain-lightning deity), who appeared to her both as a huge snake and as a man, fathering her child who also appeared both as a serpent and as a human. The narrative explains how this marriage resulted in treasure—five boxes of money—for the young woman's father (implying a bride price, standard in Mesoamerica). This story ties wealth to lightning, and lightning is connected to Rayo's cave home in the mountain, a mountain in which there is treasure, some of which was brought out of the mountain cave for the benefit of humans (the young bride's father and his wife). In much of Mesoamerica even today maize and money are both considered treasure (or wealth), so one might interpret this as an implied metaphorical narrative about Rayo bringing maize (treasure) to humans. At least indirectly we might consider this a modern reflection in Oaxaca of an earlier allegory about Rayo splitting open a mountain to reveal maize for humans.

---

[27] This is reminiscent too of the Oaxaca Chontal narrative about how the Christian God tried three times to make it rain, but then gave up, after which the mountain lord was able to finish the task (Turner 1972:59).

Mixtec speakers of Nuyoo in Oaxaca "say mountains are repositories of water; the caves, or 'houses of the Rain,' are conduits for the distribution of this water throughout the world. Caves are broadly associated with wealth; in many stories they are described as containing the valuable things in life, such as corn, beans, musical instruments, guns, and even trucks. The tenuvi, as people of the mountain are mediators of this abundance, able to call it forth for the benefit of Nuyootecos" (Monaghan 1995:349). Although the rain-lightning deity is here associated with a maize-containing mountain from which the maize is called forth by "rain people," the splitting open of the mountain by lightning lacks explicit substantiation.

To summarize, the story of lightning splitting open a mountain to reveal maize—widespread, important, and still remembered in Mexico and Guatemala—appears to have left little direct and explicit evidence in today's Oaxaca of a possible one-time presence. It is present in easily recognizable form, however, elsewhere in Mesoamerica.

Lightning: Serpent, Eagle, and Jaguar

Some data have already been presented linking lightning and the rain-lightning deity to serpents. More evidence for such an association is provided here, along with the suggestion that thunder, the noisy aspect of lightning, is linked to the jaguar in Mesoamerica as it is in Central and South America, and that there is a thunder-bird implicated as well.

As Taube indicates, "There is widespread evidence that snakes are widely identified with lightning in the Maya region both in the pre-Hispanic era and in the ethnographic present" (1992a:19). Vogt and Stuart (2005:166) maintain that for highland Mayans the co-essence relationship obtains between the Earth Owner and serpents; that is to say, a serpent is the familiar of the lightning deity. Thompson (1970) augments this highland Mayan linkage of lightning and serpents, noting first that

> Chauc, alias Anhel, is the rain god, god of water, owner of the thunderbolt, lord and owner of the mountains, protector of milpas...and intimately related to our sustenance (cf. Ah Tzenulob, "Providers of Food," as a title of the Chacs). He is lightning; he lives in the interior of a mountain, the doorway to his home being a cave guarded by a frog. That a snake is Anhel's seat can be deduced from the fact that to avoid naming a snake it may be referred to as "Anhel's seat." (1970:268)

Without referring to animal familiars, he broadens the scope of the connection between lightning and serpent both spatially and temporally by opining that "the rain cult, with world color and directional features and with quadripartite deities deriving from or fused with snakes, had developed in all its essentials in the Formative period, probably as an Olmec creation" (Thompson 1970:269). Mixe descendants of the Olmecs to this day retain traditional beliefs that a large serpent within the cave is a bridge to the abode of 'Ene· (Thunder), the lightning-rain deity (Lipp 1991:49). More explicitly, "jaguar and poisonous snakes are offspring and potencies of lightning" (Lipp 1991:29), and "each year 'Ene· sends out his 'cowboy' mounted on a jaguar, leading his offspring... to destroy the maize fields of those who have not given him offerings" (Lipp 1991:28).

Referencing Kekchi Mayans of Guatemala, Thompson says that "snakes are the servants of the Tzultacah who send them to punish man for evildoing; a light bite (nonpoisonous for a minor offense, a bite from a rattler for real depravity). In some parts, the Indians will not kill snakes for fear of angering their masters, the Tzultacah. The caves and underground homes of the Tzultacah are furnished with hammocks in which the gods rest; the cords of these are rattlesnakes" (1970:274). Here snakes are seen as servants of the rain-lightning deity or earth lord.

Iconography from the lowland Maya Classic period and the later Aztec Empire indicates that snakes were servants of the rain-lightning deities in pre-Conquest Mesoamerica. "Both Tlaloc and Chac carry snakes in their hands from time to time. These snakes are the lightning bolts they hurl from the mountains where rain gods make their retreat" (Miller and Taube 1993:150).

Taube interprets a more complex scene depicted on a lowland Maya Classic period codex style vessel, equating the serpent with part of the lightning deity's body.

> At our left a composite form of two Maya lightning gods, the Classic Chac and God K, strikes at the structure with a smoking axe and an eccentric flint. Chac appears again on our right with the same lightning weapons in a cave like hole on the roof. The composite Chac at the left has the serpent foot of God K, which coils up to cleave open the roof and penetrate the earth house. The burning serpent foot of God K is none other than lightning. (Taube 1992b:55)

Madsen informs us that in the valley of Mexico twice a year all curers go to a meeting in the cave of the rain dwarfs. Here the lightning deity is visible in his serpent transformation.

> Orders for curing are given at this meeting by the chief of the enanitos, who is called "yecacoatl" in Nahuatl and "culebra de agua" in Spanish. Yecacoatl is a big snake in charge of making rain and storms. He tells the enanitos which barrels to pour down on the earth. The Indians say that before a cloudburst the culebra de agua can be seen in the dark clouds lashing its tail back and forth. (Madsen 1955:51)

The implication is that Yecacoatl is the serpent form as well as the name of the rain-lightning deity for these Nahuatl speakers.

Totonacs also connect serpents directly to lightning. "For Totonacs the snake is a symbol of lightning, rain, and maize" (Govers 2006:308). In Nanacatlán, San Miguel is the patron of "Rain and Thunder, and therefore of agriculture. To the northern highland Totonacs and to coastal Totonacs, San Juan is the patron of thunder, rain, and water" (Govers 2006:103). For all of these it is the snake that symbolizes lightning.

The fact, mentioned above, that Zapotecs of Oaxaca associate lightning and the lightning deity with lizards and crocodiles more often than with snakes, is quite interesting in view of crocodilian behavior. It seems that the booming sounds made by crocodiles during the mating season, which is at the beginning of the rainy season, resembles the booming of thunder (Reilly 2004:7). In fact, crocodilians not only answer each other's booms but also have been heard responding with booms to the sounds of thunder. Ackerman (1988:64) notes that some alligators in Florida have been heard to bellow responses to the sounds of the space shuttle at Cape Canaveral.

Although contemporary Zapotec ethnography indicates that the lightning deity is more likely to take the form of a lizard or crocodile, Joyce indicates that in earlier times during the Olmec presence this deity was shown in Oaxaca with such serpentlike features as a bifid tongue (2009:58). Evidence of such serpentlike features can also be found at the Olmec site of Chalcatzingoin Morelos (Reilly 2004). A serpent association with the lightning deity can on occasion also be found among contemporary Zapotecs (de la Fuente 1977:266; Beam de Azcona personal communication 2011). Mixtecs of Nuyoo in Oaxaca say the rain god (*ñu'un savi*) has four different "faces" (*personas*, manifestations). He may appear as (1) lightning bolts, (2) rain serpents, which show up amid violent storms, (3) rain lizards, which appear in fine mists or dew, and (4) "rain saints," doll-like dwarfs that live in special caves called "rain houses" that are treated by people as shrines and that have little pools of water in them (Monaghan 1995:106–107). The Mixtec serpent connection with lightning bolts and the lightning deity is clear. The Mixtec lizard connection with the lightning deity accords with the Zapotec association of the two. The Mixtec region of Oaxaca is adjacent to and west of the Zapotecs. Two pictures of a rain serpent, drawn by Mixtecs, show an obvious serpent in the clouds. One of the serpents has little wings. Both have spots all over the body (Monaghan 1995:107–108), making them look like "jaguar serpents," although the spots are said by Mixtecs to represent not spots but seeds.

In Oaxaca, a definite Mixe connection between serpent and jaguar in the context of the lightning deity was mentioned above. A Mixtec connection of jaguar and serpent is also plausible in the context of a lightning deity. Mayans make a similar conflation of jaguar and serpent in at least four ways. First there is the day named Chicchan, which is the name of the Chorti rain serpent, but which has iconographic and other associations that suggest the glyph and the name for the day represent a conflation of jaguar and serpent features (Linda Schele personal communication 1992). Second, there is a linguistic ambiguity with jaguars and serpents. For example, the word *ik'bo'lay* in Chol means "black jaguar," whereas in the closely related Mayan language Cholti *ik' bo'lay* means "serpent," and in Kekchi Mayan *k'ambo'lay* means "poisonous snake" and *k'ambo'lay hix* means "margay (a miniature jaguar)." Third, images of both the serpent and the jaguar appear to symbolize co-essences or familiars. For example, one glyph for co-essence is an *ajaw* sign half-covered with a jaguar pelt with a logographic value of *way* "co-essence," and another glyph for co-essence is skeletal jaws of a serpent, also with a logographic value of *way* "co-essence" (Montgomery 2002:266). Fourth, while the jaguar is a shaman's co-essence par excellence, a connection of the shaman to the serpent is found among some Mayans and other Mesoamericans in a shaman initiation tradition requiring the initiate to sit by an anthill from which a large serpent emerges to swallow him and later to expel him through defecation, thus imbuing him with supernatural shamanic powers (de la Garza 2002:154; Corlett 1935:178; Brinton 1894:19–20). Then to balance out the often-cited relationship between the jaguar and the shaman (the shaman being a human particularly closely associated with an animal familiar), one of the words for *nagual* (co-essence) in a couple of Mayan languages have cognates that mean "serpent" in a couple of others, as noted above, suggesting a close relationship between the concepts of *nagual* and "serpent." Finally, in Tzotzil Mayan the word *chon* means both *nagual* and "snake."

According to Fitzsimmons, the main deity of the earliest people at Monte Alban was the jaguar-baby-faced god of rain and lightning who lived in a cave. This jaguar was the ancestral form of the Zapotec rain-lightning god Cocijo (1997:1), perhaps in his thunder aspect. Benson agrees that the jaguar is related to this deity and provides other suggestive linkages between the jaguar and the thunder-lightning-rain deity in Mesoamerica, both ancient and modern.

> On several pages of the Dresden Codex, the so called Rain Deity (God B) sits on a jaguar-skin seat. The Classic period Zapotec god of lightning and rain may have jaguar traits as does the deity Tlaloc in central Mexico. Such deities are more than simple rain gods; they are complex gods of fertility. Today, in Guerrero, in May, men in jaguar suits and helmets engage in ritual fights to bring rain. (Benson 1998:67; internal citations omitted; cf. Saunders 1989:161–166)

The jaguar seems a more likely candidate for symbolically (or iconically) representing the thunderous aspect of the rain deity. Many observers and indigenous communities have said that its roar, described as a loud and deep coughing noise, sounds like thunder. The serpent, with its stereotypically long undulating body, would be the more likely animal to represent lightning. In sum, this serpent-jaguar duality appears to be present in Mesoamerica, with one being emphasized more than the other in any given region and time period with respect to narrative and/or imagery.[28]

In the Valley of Mexico, the inhabitants of Teotihuacan depicted Tlaloc the lightning deity with notably large goggle eyes in addition to his putative serpent fangs and/or jaguar canines. This suggests that an owl or other bird of prey might be represented as part of Tlaloc's imagery. Eye elements are augmented with an accipitrine beak on the mask of the guardian at the entrance to the Olmec Oxtotitlan cave, this guardian possibly representing an early lightning-rain deity.[29]

A bird, depicted in a Maya codex with large goggle eyes, is also associated with lightning and the lightning-rain deity. Girard (1995:158–159) points out that Chorti Mayans equate the "storm bird" with the "storm god," noting that in accord with Chorti thought, the Madrid Codex of the Postclassic Maya depicts a bird riding on a water-laden serpent, with God B (the "storm god") substituting for the bird in some cases. The bird is called a *wak* <vac> or *wok* <huoc> bird (both are falcons) as described in the Popol Vuh of the Quiché Mayans (1995:159). Bassie-Sweet identifies the *wak* as the laughing falcon and the *wok* as the forest falcon, and further identifies the *wok* as the <muwaan bird> (or muan bird) of Classic Maya texts (2008:106–107). The laughing falcon (*wak*) eats snakes and is viewed by several Mayan groups as a harbinger of the rains as well as a healer (Bassie-Sweet 2008:142). The forest falcon (*wok*) flies faster than the laughing falcon; eats other birds, swallowing them whole; and has large owl-like eyes (Bassie-Sweet 2008:142). Falcons and eagles dive so swiftly when descending on prey that they can appear to be like a lightning bolt to both prey and observer.

In the Tzeltal tale of Ala Witz Mut—opening up the mountain of sustenance with a blast of lightning in order to reveal the maize inside—we can find another association of lightning with a bird of prey, whether it is an eagle or a falcon. Ala Witz Mut literally means "small mountaintop bird," where Witz means "mountaintop" or "peak," and he is said by some Tzeltal consultants to be an eagle (Stross 1973:104). But a small mountaintop bird of prey is also a good description of a falcon. Lipp (1991:29) reveals that for the Mixes of Oaxaca the lightning deity can assume the form of either an eagle or a serpent. The fact that Mixe *witzək* is "lightning" and *wihtzən* means "eagle" and "falcon" appears to link lightning even more closely to the eagle or falcon than to the serpent

---

[28] In the northern Andean montane and in Amazonian South America the jaguar is strongly associated with the booming sound of the thunder deity as well as with his thunderbolt, and of course with rain, while the serpent, and in particular the huge anaconda water serpent, is more closely allied with the visual effect of the lightning bolt and with sheet lightning.

[29] That an owl may be one of Tlaloc's manifestations recalls the fact that for Tenejapa Tzeltal Mayans there is a kind of owl (*mutil balam*) that seems to be tied to the jaguar. It is said that if one sees this kind of owl, a jaguar is sure to be somewhere nearby (Hunn 1977:160).

and also suggests a possible borrowing from Mixean into Tzeltal.[30] It must be said, however, that Tzeltal *witz* "mountain, mountaintop, peak" has a venerable Mayan ancestry and is semantically appropriate in the context of Sustenance Mountain. If the Tzeltal word is not borrowed, then we might conclude that the "small mountaintop bird" is a falcon and that in Mixe there is a linguistic link between the eagle (and/or falcon) and lightning.

In summary, snakes, jaguars, and eagles (or some similar predatory bird) can each lay claim to a close association with lightning and a connection to the lightning deity in the minds of indigenous Mesoamericans, past and present. Moreover, all three animals have associations with transformation and are important animal familiars, particularly with respect to war. Lightning too, we have seen, is a *nagual* for some individuals, including those engaged in war. What makes the snake, the jaguar, and the eagle stand out would appear to be their positions as the most potent predator in each of their own spheres of influence, water (or underworld), earth, and air, respectively. Some allowance must be made here for the fact that it is not clear which snake is being referenced. The tropical rattlesnake is large and deadly and so is the fer-de-lance, but neither is particularly associated with water; nor is the boa constrictor. The huge water boa, or anaconda, would fit here perfectly, but it is native to South America and not present in Mesoamerica. Moreover there may be a question about whether the falcon or the eagle better represents lightning in the sky. At least with the jaguar there seems to be no competing icon.

Lightning, Twins, and Triads

According to Frazer (2000:262–269) the belief that twins have magical powers over the weather, and especially rain, is widespread in the world. His examples from the Pacific Northwest Coast, Australia, Africa, and South America illustrate this contention. A Mesoamerican connection between lightning or rain and twins exists, but more on the basis of inference than direct associations.

For example, Nuttall, in discussing ancient "superstitions" of highland Mexico, attributes to sixteenth-century chronicler Gerónimo de Mendieta the information that in Nahuatl, "the name for twins was *cocoua*, which is also the

[30] Several different birds are associated with rain in various parts of the world. The eagle dance found in some native North American groups is in essence a rain ceremony (Marder 2005:138). The Assyrian storm deity had an eagle as his familiar, symbolizing lightning and rain. In Mesoamerica Tzeltal Mayans and Mixes associate the woodpecker with lightning, similar to European mythology, in which the woodpecker is associated with lightning and fire. Totonacs (Govers 2006:160–163) attribute to the woodpecker a role in trying to get maize from Sustenance Mountain similar to that of the woodpecker in Tzeltal and Mixe narratives, and in the end, for the Totonacs it was the vulture that received the lion's share of the maize.

name for serpents. According to an ancient tradition the first woman who bore twins was named Coatl or Serpent, and therefore twins were also named serpents. When both were allowed to live, one of them surely killed or devoured one of its parents" (1897:276). López Austin adds that twinned serpents, composing the skirt of the Aztec goddess Coatlicue (Serpent Skirt), as depicted on a monolith, represent fertility and fecundity (1997:236). Related information is that the Aztec goddess Cihuacoatl (Serpent Woman), another appelation for Coatlicue, named and portrayed visually as a serpent goddess, was both mother of twins and mother of the human race, again relating twins to the serpent—a co-essence of the rain and lightning diety—but not directly to lightning or the lightning deity. It is because we already know of the strong linkage between the rain-lightning deity and serpents that the association of serpents with twins becomes then a plausible inferred association also of lightning with twins. Snakes, relatively well known to have bifid tongues, are also relatively frequently found in nature with two heads, as compared to other animals, and both facts may be related to an association made between twins and snakes.

In the sacred Quiché Mayan narrative known as the Popol Vuh, written in the Spanish alphabet and translated into Spanish during early colonial times, a pair of male "hero twins," along with their half brothers, the "monkey twins," are probably the best-known twins in Mesoamerica. In this story the "hero twins" end up as the sun and the moon, and they are associated not only with the sun and the moon but also with maize, the Mesoamerican ballgame, and the slaying of monsters; but there seems to be no direct connection of either twin pair with lightning or with rain or even with snakes. In an important Mixe origin myth orphaned twins, a brother and sister, ultimately become the sun and the moon, and they have adventures distinct from but reminiscent of the Popol Vuh's account of the hero twins. The Mixe twins sleep in a maize granary, shoot arrows at a monster in a tree on a mountain, summon a toad to carry a message, and so on (Lipp 1991:75–76). Among other things, the sister urinates from the pinnacle of the mountain, thereby saving humanity. As mentioned below, toads and frogs can be related to lightning and to the lightning deity, and as Lipp points out "other Mixe texts emphasize urinating as a fertilizing act, urine being analogous to water and rain" (1991:76).[31] Moreover the arrows shot at the monster, like the urine rained down from the peak, might

[31] Significantly, the three main nutrients of plants, including maize, are nitrogen, phosphorus, and potassium in absorbable form, and urine contains large amounts of nitrogen and significant amounts of the latter two. Used as fertilizer, urine is optimally diluted by about 10:1 with water to avoid burning plant roots, and while it is well known to be currently used in this way in China, one suspects that urine has been used in just such a way in many parts of the world since the advent of agriculture if not earlier. While lightning "fixes" nitrogen electrically (breaking molecular bonds to allow recombination), urine provides nitrogen biologically; both make nitrogen available for plant use, and so each could be conceived as apposite symbols of the other.

seem to be metaphors for lightning bolts, so there may be indirect connections suggested in this narrative between twins and lightning. However, as in the Mayan Popol Vuh, the Mixe narrative directly associates twins with maize, the slaying of monsters, and the sun and the moon, but nothing links twins directly to rain or lightning (Lipp 1991:117).

Another apparent association of lightning with duality, though not necessarily with twins, comes from Cancuc Tzeltal speakers who say there are exactly two kinds of lightning bolts: green lightning bolts (*yaxal chauk*) and red lightning bolts (*tzajal chauk*) The green lightning bolts belong to the mountain lords, are located inside important mountains, and emerge from the mouths of caves with rain clouds. They are relatively weak, as evidenced by the lapse of time between the flash and the sound. The red lightning bolts, co-essences or "spirit companions" of living people, are strong. The latter, during the dry season, remain dormant in the moss and orchids of tall trees in the tropical forest (Pitarch 1998:216). When they become active at the start of the rainy season, their job is to battle the storm winds that are trying to destroy the *milpa* by knocking over the maize stalks.

Although twins are related to rain in many parts of the world, including societies located both north and south of Mesoamerica, the sparse evidence for similar traditions in Mesoamerica exists only indirectly and by inference from Aztec, Mayan, and Mixe data, nowhere being directly encountered in Mesoamerica, despite the obvious potential for such an association in the fact that thunder and lightning, both clearly related to rain, occur roughly simultaneously and appear to be different manifestations of essentially the same phenomenon, and therefore symbolic twins.

In Mesoamerica lightning can be more directly associated with a triadic pattern. For example, Pinola Tzeltal speakers retain a triad of deities: Whirlwind (Sutu ik'i), Lightning Bolt (Chauk), and Meteor or Fireball (Paslam) (Hermitte 1964). The Popol Vuh, written down in colonial times, distinguishes among three types of lightning that together constitute "Heart of Sky." The first one is called Thunderbolt Huracan (which we can view as akin to the whirlwind). The second one is Youngest Thunderbolt, and third is named Sudden Thunderbolt (Christenson 2007:60). Bassie-Sweet (2008:102–121) views this as a reflection of a lowland Classic Maya triad that she identifies as the Thunderbolt Triad (2008:102), consisting of three brothers, always named in the same order, eldest, youngest, and middle. She identifies GI of the Palenque triad of gods as Thunderbolt Huracan. GII (generally seen as K'awiil or God K of the codices) is thus the Classic period equivalent of Youngest Thunderbolt from the Popol Vuh, and GIII is identified with Sudden Thunderbolt (the weakest of the three). She further interprets the triad of Palenque lightning deities as manifestations of the three stones of Maya creation as revealed in Classic Maya script from Quirigua in Guatemala. These stones are arranged in a triangular pattern of "throne-stones," as seen in the constellation Orion and symbolized by the three stones of the prototypical Mesoamerican hearth here on earth (Bassie-Sweet 2008:121; Carrasco 2010). She concludes that the "striking parallels between the three Heart of Sky thunderbolt gods and the Palenque triad indicates that these deities represented core beliefs about gods of storms, lightning, and fire and their role in creation" (Bassie-Sweet 2008:121). These identifications suggest conceptual continuity from at least Classic times through to the present of a triad of storm deities representing thunder and lightning and associated with rain.

The pattern of three that is associated with lightning and the lightning deity during Classic times among the Maya is also found in the Maya glyphic script, in which the head variant glyph for number three shows that this number is ruled by the lightning god. As Thompson (1971) has demonstrated, the day-names from Caban to Muluc are aligned with the numbers one through thirteen in that the ruler of number one, the moon goddess, also rules the day Caban, just as the ruler of the number eight and the day Kan is ruled by the maize deity, and, significantly for our purposes, the ruler deity of the number three is the rain-lightning god, who also rules the day Cauac. Cauac means "rainstorm" or "lightning" and corresponds to the Aztec day named Quiahuitl "rain."[32]

In central Mexico, Florescano (1999), looking at Teotihuacan iconography, suggested a triad of agricultural deities: Goddess of the Cave, representing fertility; Tlaloc, the rain-lightning deity; and the Feathered Serpent, representing vegetation renewal and related to Venus, associated with thunderstorms, the rainy season, and war by Aztecs and by the Classic Maya. We have already seen how closely linked symbolically the cave and the serpent are to the rain-lightning deity, the former being his abode, the latter being his animal spirit companion and guardian of his abode. The "feathered" aspect of the Feathered Serpent (Quetzalcoatl) might in this context be seen as relating to the eagle or the falcon (though in other contexts the quetzal is seen as the bird of relevance to the name in Aztec perceptions).

In the Miahuatlan Zapotec-speaking town of San Bartolome Loxicha of the southern Oaxaca mountains, Mdi', the lightning deity, is offered special triangle-shaped *memelas* (similar to fresh maize tortillas, but a bit thicker), without salt and topped with beans, for the blessing of the milpa ritual. The three corners of one of the triangle shaped *memelas* are broken off and placed, along with the smallest *memela* and other offerings, in the sacrificial hole in

[32] Above we touched on the similarities in sound, meaning, and traditional associations between the Guatemalan highland name for the eighth day (counted from Imix as the first), Q'anil, and the Yucatec name, K'an, for the day ruled by the deity that also rules the number eight (eighth as counted from Caban as the first). We might therefore compare the Quiché third day (counting from Imix as day one), which is Aq'abal, with the Yucatec third day, counting from Caban as day one, i.e., Cauac, meaning "lightning" and "rainstorm." Significantly, a child born on Aq'abal acquires an extra soul called "lightning" (*coyopa*) and might well become a "priest-shaman" (Tedlock 1992:110).

the milpa (Rosemary Beam de Azcona personal communication 2011).[33] In nearby towns of Santa María and San Vicente Lachixío, Zapotec speakers fashion triangular tamales that also have an apparent connection to the lightning deity according to field notes of Mark Sicoli, who suggests the possibility of a symbolic link between triangles and the angles of lightning bolts (personal communication 2011).[34] Triqui speakers of Chicahuaxtla, Oaxaca, say that a triangle of ribbons on the Triqui *huipil* (female blouse or dress) symbolizes rain and the rain serpent (Fernandez 2006:36), and triangular cuts on ribbon edging that cover seams symbolize the "little feet of [the rain] god." Evidence from Precolumbian Oaxaca indicates that similar triadic associations with lightning go back considerably in time (Urcid 2005:139, 140, 142).

Likewise triadic associations in the Mayan region extend from the present back through colonial times at least until the Maya Classic and perhaps beyond. Quiché speakers today single out an almost equilateral triangle-shaped group of stars in the constellation we know as Orion (a triangle composed of Alnitak, Saiph, and Rigel), maintaining that these are the heavenly reflection of the three hearthstones in the Quiché home. Centrally located within the heavenly triangle, the cloudy appearance of nebula M42 resembles smoke from a fire (B.Tedlock 1992:181–182; D. Tedlock 1996:236). In one colonial Yucatec document, the Chilam Balam of Chumayel, a maize spirit referenced as the "three-cornered jade stone" remained within maize, prior to Creation, emerging at a later time with long locks of hair (Roys 1967: 107; Thompson 1970: 351). This triangular stone clearly refers to the maize ear and in Classic period Maya imagery was represented as sometimes pointed pendant jadeite celts, typically in groups of three (Taube 1996a:42). The reference to the the triangular stone as the maize spirit at the time of Creation invites comparison with what Freidel, Schele, and Parker (1992:69, 79) refer to as the triangular "Hearth of Creation" or "First-Three-Stone-Place" identified by them in the inscriptions from

---

[33] Rosemary Beam de Azcona's fieldwork and careful observation in the southern mountain region of Oaxaca has led her to associate the Zapotec lightning god/rain deity with "reptilian animals, triangular shaped food, and a distaste for salt" (personal communication 2011). Since the Mesoamerican three-stone hearth is in a triangle shape, is used for cooking, and contains fire, a "relative" of lightning, triangular foods may have an iconic and indexical motivation on which the symbolism is built. With respect to salt, "in stories collected from San Baltazar Loxicha, the preference for unsalted food is often used as a type of foreshadowing, indicating that the supernatural, particularly Lightning, is involved. For example, a mysterious stranger will appear in town, begging for food, but refusing salt, and later this person always is interpreted as having been not truly human. In San Bartolome Loxicha the food that is prepared for Lightning, when sacrifices/offerings are made, is prepared without salt" (Rosemary Beam de Azcona personal communication 2011). It may be noted that salt is an excellent conductor of electricity and salt water as well as soil high in salt appears to attract lightning. Thus avoiding salt when making offerings to the lightning deity would seem to make good protective sense.

[34] A stylized arrowhead is also shaped like a triangle, and lightning bolts have been likened to arrows in Mesoamerica, as have serpents, whose heads also bear resemblance to both triangles and arrowheads, further reinforcing the lightning-triangle link.

Palenque and at Quirigua as referring to the Orion triangle around M42, that is to say, the Quiché hearthstones. In the Zapotec triangular *memela* made of maize we might also see a spatially displaced reflection of the triangular, triadic maize representative, ritually re-creating the original maize of Creation, and in so doing the *memela*'s connection to lightning might in turn hark back in Zapotec ritual tradition to the time of Creation and its place in the Orion constellation.

In sum, the twin connection with lightning and rain is implied and indirect in Mesoamerica, and it is mentioned here in connection with the well-documented relationship between lightning and twins in the Andes (as well as in other parts of the world) primarily for comparative purposes. The better-documented relation of three to lightning in Mesoamerica shows up in triadic classification of lightning deities, glyphic accord between the rain deity and god of number three, and triangles symbolically related to lightning. Here the association is clearer and more directly made.

Lightning Strikes and Thunderstones

Almost everywhere in Mesoamerica, flint and obsidian chips, flakes, and cores as well as jade axes are viewed as products and symbols of lightning and the lightning-rain deity (**Figure** 42). Even beyond the crashing thunderlike sounds produced by such stones in their usual applications, the symbolism is obvious when we recall that some stones when struck together, and particularly when flint or other siliceous stones strike iron (in the form of magnetite, pyrite, hematite, ilmenite, or meteoric iron), can produce sparks that resemble miniature flashes of lightning, and like lightning, these sparks can produce fire. Additionally lightning can be seen to strike earth with great force, sometimes chopping a tree in two as might a stone axe. To top it off, lightning can produce hard stone-like entities called fulgurites when it strikes silica derived sand or soil. Some of the stone artifacts seen as products of lightning strikes are tools of agriculture, while many are tools of warfare, which, as we have seen above, has its own direct connection to lightning and the lightning deity. Through nomenclature Mesoamerican peoples of today distinguish between flint, obsidian, and jade, just as did their ancestors in earlier times. Flint is more often linguistically related to fire and sparks, while obsidian is more frequently associated linguistically with thunder and the lightning deity. Obsidian and jade appear to share Mayan associations with the color green.[35] Chol Mayans today sometimes refer to obsidian as *y-ehk'ach Lak Mam*, literally "fingernails of Our Grandfather," the

[35] Green obsidian found in the Maya region is sometimes seen as both special and diagnostic of having probably come originally from faraway Pachuca (north of present-day Mexico City).

FIGURE 42 Various stone artifacts, including several greenstone celts. All would be considered by many indigenous Mesoamericans to be products of lightning strikes. Lightning strikes are also thought to produce worked pieces of flint or other siliceous stone. (Photograph courtesy of David Menz)

latter being an epithet for the lightning deity or mountain lord (Linda Schele personal communication 1992; see also Schele and Freidel 1990:463). Coatlán Mixe speakers call obsidian *enee-dzaa* "thunderbolt stone." Similarly, Ixcatlán Mixes call obsidian "thunderbolt blade" (Lipp 1991:29:217), as do Camotlán Mixes (Miller 1956:246).

Aztecs at the time of the Spanish Invasion compared the flint-tipped hurtling lance and arrow to a stroke of lightning. The Postclassic Aztec rain deity, Tlaloc, was depicted in the codices wielding a jade axe said to symbolize a thunderbolt, just as the Postclassic Maya deity known as Chac (or Chaak) was. Recinos tells us that the black volcanic stone we know as obsidian, *chay* in Quiché, is "the 'stone of lightning' of the peasants; from it the Indians selected small sharp pieces which they used as knives, razors, and arrow points" (Recinos et al. 1950:117 n. 3), and to the Quiché Mayans the day named Etz'nab with a traditional meaning of flint knife is associated with lightning (Carmack 1981:85). The corresponding Aztec day-name, Tecpatl, means "flint" and likewise symbolized lightning. In today's Nahuatl of the Huasteca region in Mexico, thunder is called *tlatomoni* and lightning is called *tlapetlani*, and both have spirits linked to the "rain dwarfs" responsible for the rain that is brought into being when they strike together the implements they carry, creating thunder and lightning displays that come with the rain (Sandstrom 1991:247, 250). A Nahuatl speaker in the Huasteca region, when asked about flint chips found next to pathways, responded that they were sent to earth by lightning strikes (Stross field notes 1971).

"Ethnographic sources indicate that lightning bolts were thought to be flint axes thrown by the Chahks" (Bassie-Sweet 2008:7), and "lightning bolts and meteors are associated with obsidian and flint, and both are considered to be weapons of the thunderbolt gods" (Bassie-Sweet 2008:120). Obsidian pieces are to Chol Mayan speakers "axes of our grandfather" (i.e., the lightning deity) because lightning is thought to have created it (Aulie and Aulie 1978:61). Spero (1991:188) provides other references for Mayan groups believing stone axes to be the axes of the lightning deity. A Huastec Mayan variant of this theme is that obsidian chips found in the paths are believed to be pieces of the sky downed by the lightning deity when using his axe (Alcorn 1984:141). Tedlock says that "children of the thunderbolt" for the Quiché Mayans include the glassy fulgurites that are produced by lightning striking the earth (1996:224).

According to Chorti Mayans, "lightning-makers," servants of the lightning deity, carry stone axes with which they beat the rain out of the clouds. Sometimes they will throw one of these flint axes (*tok' tun*) to earth trying to kill a Chicchan (lightning serpent) (Wisdom 1940:396, 673). Chorti speakers call stone axes "cloud stones," as they do the flint from which many of them are sourced, also thus "associated with the rain-making deities" (Wisdom 1940:382). Stone axes are also called "fire stones" by Chortis, who believe that wherever lightning strikes, a stone axe can be found buried in the ground (Wisdom 1940:382). Mopan Mayans also believe obsidian and flint celts to be the lightning deity's axes hurled at the earth (Baudez 1992:45). In looking to lightning as a possible source for the "year sign" worn by the deity of lightning and rain in Oaxaca and in central Mexico, Fox and Justeson say "Campbell notes that obsidian and flint are referred to as 'lightning shit' in southern Mesoamerica" (1984:72; internal citation omitted).

Mixes of Oaxaca are documented as having similar beliefs about obsidian blades and chips coming from where lightning has struck the earth, or perhaps broken off from the machetes that the *rayos* (lightning deities) employ (Miller 1956:246; Lipp 1991:29, 217; Hoogshagen and Halloran 1993:58). Mixtecs, also of Oaxaca, think of obsidian as "fossilized lightning" (Monaghan 1995:109).

It is clear, then, that in addition to the strike of a lightning bolt having the power to induce knowledge (as in shamanic initiation and practice), to deliver and fertilize maize (as in splitting open Sustenance Mountain, "fixing nitrogen," and germinating maize), and to grow mushrooms (to be discussed below), it can also create stone tools and weapons of war according to Mesoamerican traditions. To provide context for these Mesoamerican links of lightning to stone artifacts it must be noted that in Europe stone celts have also long been seen as products of lightning striking the earth, and in some cases the names of the lightning-created stone celts can be translated literally as "thunder stones." The ubiquity of worked stone implements and celts being called "thunderstones" and similar terms in Mesoamerica as well as in the Andes taken in conjunction with the observationally logical reasoning that could lead to such nomenclature makes it clear that independent invention rather than diffusion by loan translation explains its incidence in the three different regions.[36]

Lightning and Tobacco

Tobacco (*Nicotiana tabacum* L. and *N. rustica* L., Solanaceae)[37] is almost everywhere in Mesoamerica viewed as a potent means of warding off undesirable forces, such as witchcraft, spirits of the dead, snakes, worms, centipedes, scorpions, and lightning. It is currently taken through the mouth in two different forms: rolled up and smoked as cigars or cigarettes, or powdered and carried with lime (calcium hydroxide) or ash (potassium hydroxide) in a gourd container to be placed in small amounts in the mouth and kept in the cheek.

Witchcraft is undesirable and difficult to protect oneself from, and protection against witchcraft is among the several desirable effects of tobacco. Tzeltal Mayans along with Tzotzil speakers believe that rubbing the joints, or even passing tobacco over the joints, can protect one from witchcraft. Similarly, among the "Mazatec of northern Oaxaca . . . a shaman starts to rub the forearm of a pregnant woman with ground tobacco and lime a month before parturition to make her invulnerable to witchcraft. . . . Their neighbors, the Chinantec, also used tobacco to ward off witchcraft" (Thompson 1970: 121). Mixtecs of Oaxaca also protect themselves from witchcraft with tobacco (Monaghan 1995:350). Since shamans (and other curers) are the most likely individuals to be accused of witchcraft in Mesoamerica, this is really another connection between tobacco and shamans, who use it in curing various ailments, including witchcraft caused by other shamans, as well as in making rain. Thompson (1970:118–120) lists many illnesses and ailments along with bites, stings, body invasions, and other complaints all of which are treated with tobacco in central and southeastern Mesoamerica, whether by being ingested, rubbed on, or

---

[36] When a word or phrase is borrowed from a language by literally translating the root form(s) it is called a loan translation or calque. In Mesoamerica there are a number of such forms.

[37] *Nicotiana rustica*, a wild species of tobacco has yellow flowers and is up to three times as potent as *Nicotiana tabacum*, the cultivated species, which has flowers that are white to pinkish in color. *N. rustica* is used alone or in combination with other botanicals by various peoples from western Mexico to South America to induce altered states of consciousness (Wilbert 1987). In much of Mexico *N. rustica* is called in Spanish *piciete*, after the Nahuatl word for wild tobacco.

blown, sprayed, or spit from the mouth of a shaman. In addition to warding off witchcraft, it can serve as a vehicle for transmitting it.

Smoking wild tobacco (*Nicotiana rustica*, L.) to induce a shamanic trance, called tobacco shamanism, is a practice of some South American indigenous groups like the Warao, described in detail by Wilbert (1987). Similar uses in today's Mesoamerica, if they exist, have not been described, but some scholars suspect that tobacco may have been so employed by the Classic Maya. Classic period iconography illustrates cigar smoking, which is suggestive. Thompson, referring to indigenous Mesoamerican tobacco practices during the Colonial period (approximately 1521–1821), claims that "it was as an inducement to trance conditions that tobacco appears to have functioned more often in divination" (1970:115). In the Mixtec-speaking town of Nuyoo, certain humans, called *tenuvi*, "rain people" or "mountain people" by the inhabitants, and "rain shamans" by some scholars, are associated with lightning and can bring the rain. One way they do this is by ascending four mountain peaks on which rain shrines are located and smoking seven cigars made from locally grown wild tobacco. Smoking these cigars, it is said, causes rain clouds to form and inebriates the *tenuvi*, causing him to shout thunderously, with rain as a result (Monaghan 1995:347–350). This is peripheral evidence that others in Mesoamerica, including Classic Mayas, may have used it simililarly, with resulting dissociative states.

Tzotzil Mayans, like many Mesoamericans, use tobacco for protection and for curing several ailments. They also have great faith in tobacco's ability to repel both lightning and serpents (**Figure** 43):

> When rubbed on the body, it protects one from lightning strikes. A man pursued by a sickness-causing rainbow can neutralize it by scattering tobacco powder on the ground (Breedlove and Laughlin 1993:243). Similarly, tobacco juice spat toward an approaching storm will divert the winds, driving away the horned serpent (*xulub chon*) said to ride inside these destructive tempests. Indeed, spitting tobacco juice at snakes of all types—said to be the "dogs" of the *anjeletik*—can paralyze them; and should one be bitten, a paste of chewing tobacco neutralizes the venom. (Groark 2010:18)

Guiteras-Holmes was advised by a Tzotzil healer from San Pedro Chenalhó that tobacco is equivalent in some respects to lightning and the lightning deity, named Anhel (the Earth Owner): "Lightning is 'anhel, and to avoid being killed by it, the people use moy, which is their protector" (1961:235). Also, "Moy [tobacco] is 'anhel and it is placed at the head of the bed in a very small gourd bottle when one is ill" (Guiteras-Holmes 1961:177). *Moy* has other uses as well: " 'On hearing thunder the people bring out their moy [powdered tobacco] and keep it in their cheek, and in this way it will not thunder too loudly. When we die the pilico [moy] defends us: we will rest.... When someone is ill and a

FIGURE 43 Native tobacco (*Nicotiana rustica* L.). Mesoamerican traditions, like those of many other indigenous peoples in the New World, hold that tobacco, including *Nicotiana rustica*, will protect from or ward off snakes and thunderstorms. The plant grows three feet tall, with yellow flowers and foot-long leaves, and is significantly more potent than commercial tobacco made from the species *Nicotiana tabacum* L. (Public Domain, Wikimedia Commons—image after Gustav Pabst, *Kohler's Medizinal-Pflanzen in naturgetreuen Abbildungen und kurtz erlaüterndem Texte* [Köhler: Gera-untermhaus, 1897])

companion dies in the same house or in a neighboring one, pilico [native tobacco] is rubbed on all his joints, on his stomach, his head, and the little gourd-bottle is placed at the head of his bed' " (Guiteras-Holmes 1961:217–218). Tzotzil speakers in San Pedro Chenalhó not only speak of both lightning and tobacco as Anhel, directly connecting tobacco and lightning (Guiteras Holmes 1961:209, 217, 235, 264; Thompson 1970:116); notably, they also refer to the spirit of maize (*x'ob*) as Anhel (Guiteras-Holmes 1961:218).[38] The Anhel (Earth Owner, lightning deity) is of course also closely connected with snakes whether they function as his animal familiar, his "seat," or his assistants.

The linkage of lightning and snakes to tobacco provides additional evidence that the three are associated with one another, as has been demonstrated above. Other Mayans, including Huastecs, also think that tobacco can ward off snakes, and Totonacs in northern Mesoamerica share this belief (Winter 2001:54, 55). Likewise, the Aztecs saw tobacco as a way to ward off or kill snakes (Thompson 1970:119–120). Mixes chew tobacco in order to suck the venom from a snakebite wound (Lipp 1991:185), and Huastecs also treat snakebite with tobacco (Alcorn 1984:712).

Tobacco has similar associations in North America. Navajos smoke tobacco during hunting trips to prevent injury from lightning and snakes (Winter 2001:270, 275), and Kickapoos of northern Mexico offer tobacco to "the Thunderers" to dispel approaching thunderstorms (Winter 2001:19). Tobacco can also ward off approaching thunderstorms with lightning according to the Seminoles of Florida (Greenlee 1944:323), a belief shared on many of the Caribbean islands (Winter 2001:57).

It would make sense for tobacco to protect against lightning, when it is viewed as protection against snakes, because lightning and snakes are co-essences in Mesoamerica, as noted above.[39] The widespread connection between lightning and snakes and lightning and tobacco may also be understood in the context of the fact that snakes actually are repelled, and can be killed, by tobacco, as a noted herpetologist once told Elaine Schele (private communication 2006). Perhaps the fruits of good observation of tobacco and snakes led peoples in diverse regions to extend tobacco's effects on snakes to its effects on other things associated with snakes. Recent research also suggests an as yet unexplained connection between lightning and tobacco. That is, lightning-caused fire appears to encourage the growth of at least one species of tobacco, *Nicotiana attenuata* (Adams and Toll 2000:144).

---

[38] Lightning has a close association with maize, as noted above in connection with the lightning deity's relation to maize fertility and to his act of splitting open the mountain with maize inside so as to make maize available to humans. As some put it, "Thunderbolt is the owner of corn" (Bassie-Sweet 2008:106).

[39] It would be a bit more accurate to say that snakes and lightning are co-essences of the lightning deity.

A final association of tobacco with lightning, indirect but relevant, is found in the Aztec festival to Huitzilopochtli, the war deity. Held on the day named Flint Knife (1 Tecpatl), it featured burning tubes of tobacco (Thompson 1970:113). War and flint knives, closely connected to each other, have both been seen already to relate to lightning, and here we see an association of both with tobacco in ceremonial activities.

Thompson views the chewing of tobacco and lime, ashes, or some other form of powdered alkali that reacts chemically with the tobacco to make available its desired stimulant properties, as a "transmutation of the Andean habit of chewing coca leaves with lime. In both areas the mixture, or the lime alone, is stored in narrow-necked gourds and dipped by means of a spatula" (1970:112).

In summary, tobacco has multiple connections with lightning, snakes, and shamanism, sharing in a pattern of associations that maintain an ideological complex of beliefs that can at least in part be related to observations in nature. For example, we know that (the nicotine in) tobacco is an effective insecticide, snake repellant, and medicine for humans, essentially ridding the body and the garden of unwanted pests and ailments. It also can be used to put shamans and others into a trance state, or if used in less potent forms and quantities it can have desirable effects on thought and feeling. These observations appear to explain in a natural way why one might connect tobacco with snakes and with curers (shamans). Since both are already associated with lightning, it is but a small step to make the additional connection. That is, if tobacco can ward off snakes, which are sometimes co-essences of lightning, then it is a simple matter of metaphorical extension to see it as warding off lightning.

Lightning and Mushrooms

It is a common observation that mushrooms often spring up literally overnight after a soaking from a lightning storm, an observation that Mesoamericans likely made very early on. Thus mushrooms can be related to lightning by folk observation that the more lightning there is, the more mushrooms spring up, and to the occasionally encountered belief that lightning may even be responsible for creating mushrooms.[40]

We may recall here that lightning delivers one form of nitrogen fixation, which is necessary for growing plants. Even warfare, when believed to cause rain by dint of loud sounds, could thus be seen as related to plant fertility as well as mushroom growth because of the din resembling thunderbolts and lightning strikes. Though mushrooms are no longer considered

[40] Stross field notes on Tenejapa Tzeltal beliefs 1971.

plants according to scientific biological classification, they appear to share in some of the beneficial effects that lightning and rainstorms are known by both folk observations and modern science to have on plants.[41]

The Mesoamerican rain deity's preference for smaller humans, usually in the form of children, has been mentioned above in the context of sacrificial offerings, and this preference can be amplified here in a nonsacrificial context; that of the Yucatec Maya rain ceremony, called ⟨*ch'a chaac*⟩ "bring rain." This ceremony makes use of four frog impersonators, usually young boys, who are tied to each of the altar's four posts (Redfield and Villa Rojas 1932:142). A similar rain ceremony was until recently also undertaken by Chortí Mayans (Girard 1995:136). These ceremonies link children to frogs, as discussed below, not only showing that both are related to rain and therefore lightning, but also introducing the following links between children and mushrooms.

The Mazatec shaman María Sabina of Huautla, Oaxaca called mushrooms "the children" and referred to the psychoactive mushrooms used in her shamanic rituals as "the holy children." Wasson (1980: 122–134) devotes several pages to linguistic and iconographic evidence that Mazatecs, Mixtecs, and Chinantecs of Oaxaca and Nahuatl speakers from Santa María Tonantzintla, in the Mexican state of Puebla, all think of, and refer to, mushrooms as children, whether in native terms or by using the Spanish words *niños, chamacos,* or *doncellas.*

Two Mixe speakers in Ayutla, Oaxaca, told me that mushrooms, and especially psychoactive mushrooms, are often referred to as children, for they have just been "born from the earth." In Camotlán, Miller (1956:219) heard that on ingesting hallucinogenic mushrooms sleep is always followed by the same vision—the appearance of two dwarfs (or little ones), a boy and a girl who talk with the sleeping one and answer his questions. In the southern Oaxaca highlands, the Mixes around San Juan Ixcatlán speak about the offspring of Thunder ('Ene·), which "take the shape of snakes or small, naked boys.... When they become adults they take on the 'burden' of the thunder god" (Lipp 1991:35). These children of Thunder appear in mushroom-induced psychotropic visions as children (Lipp 1991:36). Mixe speakers call the hallucinogenic "mushrooms of the earth" (*Psilocybe mexicana* Kumm) *'ene· di:z* "Thunder's teeth" (Lipp

[41] In 2010, Japanese researchers confirmed the accuracy of Japanese agricultural folklore maintaining that good mushroom harvests come when there is an abundance of lightning, by experiments with eight different types of mushroom. The harvest of shiitake mushrooms, for example, was doubled with the application of artificial lightning (Ryall 2010). The explanation for such dramatic effects has not yet been established, and there are a number of competing suggestions. Some say it is the ozone produced by the lightning; others that it is the result of nitrogen fixation and subsequent watering with nitrogen in usable form. The Japanese scientists are considering the possibility that exposure to danger is what produces the bumper crops of mushrooms, based on a need to reproduce in response to the possibility of death by direct lightning hits. It is, after all, the fruiting bodies, the reproducing form of the fungus, that are harvested for eating.

1991: 29). Given these multiple linkages it is difficult not to see the rain-lightning god and his dwarf servants as connected quite closely both to mushrooms and to children. A further linkage can be found between mushrooms and maize in that Mixe speakers believe that the "growth of mushrooms is closely associated with that of maize, in that the mushrooms, 'planted' by termites[,] . . . grow when the maize ears are ripening in June and become 'mute' and dangerous when the maize is tasseling during the dry, hot spell" (Lipp 1991:189).

Another psychoactive mushroom, the *Amanita muscaria* or "fly agaric," is called "thunderbolt mushroom" by Quiché Mayans, a fact noted long ago by a mycologist named Lowy (**Figure** 44). Tedlock "can confirm Lowy's report that the Quiché word for thunderbolt, *kaqulja*, is also the term for the *Amanita muscaria* mushroom (Lowy 1974:189). . . . The names Newborn and Sudden Thunderbolt suggest the rapidity of mushroom growth. Literal bolts of lightning are sudden as well" (1996:294). One can relate the name Thunder(bolt) in words for mushroom to the fact that the *Amanita muscaria* is psychoactive and used by some shamans in ritual activities and communication with the gods, as notably practiced among Mazatecs and Mixes, who are themselves

FIGURE 44 *Amanita muscaria*, the "thunderbolt" mushroom of the Quiché Mayans of Guatemala. It is psychoactive, like the *Psilocybe* mushrooms used by Mixe curers of Oaxaca, Mexico, who refer to these mushrooms as "thunderbolt." (Licensed under the Creative Commons Attribution–Share Alike 3.0 Netherlands License: attribution Onderwijsgek at nl.wikipedia)

more frequently associated with other species of psychoactive mushrooms, particularly *Psilocybe* and *Conocybe* species, recalling Mixe reference to them as "Thunder's teeth." Tedlock further relates mushrooms to lightning with a line from the Popol Vuh. "Nanahuatl means 'warts' in Nahuatl (AM) giving us one of several bits of evidence that there might be an allusion to mushrooms in the names of the previous line, 'Newborn Thunderbolt,' 'Sudden Thunderbolt'" (1996:294). This same Aztec Nanahuatl is the Aztec deity who threw thunderbolts to break open the mountain containing the first maize.

It seems that Mesoamerican lore about lightning's positive effect on crops and mushrooms parallels similar but surely independently derived agricultural lore in Japan. Edible mushrooms, like maize, are an important form of treasure or wealth for people, and such treasure as might be called "children of lightning" constitutes a strong association of mushrooms with lightning (not to mention with maize), and perhaps also an association of children with rain, helping to explain Aztec sacrifice of children to the rain deity during drought years.

Mushrooms are clearly associated as well with shamans and shamanic practice, for it is psychoactive (or hallucinogenic) mushrooms that bear the words "thunder" and "lightning" in their names. These mushrooms that seem to spring up overnight with lightning speed, that may appear to be "created" by lightning, and that can cause a form of "trance," enhancing shamanic communication with the Other World to facilitate curing and finding lost items, must be powerful indeed, as is lightning. They can alter one's perception of "reality," effecting thus a transformation. Such communication, knowledge, and transformation as may be attributed to deriving from ingestion of specific mushrooms would constitute an excellent reason for naming such mushrooms after lightning or thunderbolts.

Lightning and Frogs

Lightning has important frog associations, seemingly derived from the indexical relationship between frogs and the rain that accompanies lightning. That is to say, after a rain one notices in Mesoamerica, as in other parts of the world, that living frogs come out in abundance, seemingly out of nowhere; and they hop around, apparently with great gusto, making their appearance all the more noticeable, and associable with rain, as if they may truly be the lightning deity's "pets."

> Frogs, because their croaking announces the rains, have an intimate relation to the Chacs. They are the Chacs' musicians and guests.... The Palencano-Chol legend already cited tells us that the toads were the children of Chac's wife.... The special pets of the Chacs are the little black frogs with an orange line down their backs known as *uo*

> (*Rhynophrynis dorsalis*), who also gave their name to the second twenty-day month in the Yucatec calendar. (Thompson 1970:258)

After a thunderstorm and heavy rain another phenomenon has been observed more frequently than one might suppose, and in many parts of the world: frogs raining from the sky, often dead but sometimes still living, their bodies littering the ground. In fact, frogs are the second most common kind of organism, after fish, that are reported to rain from the sky (McAtee 1917:21). The lightning accompanying such storms, already intimately associated with rain in the person of the lightning-rain deity in Mesoamerica, could easily be viewed as the cause of the raining frogs.

There is more evidence that Mesoamericans associate frogs (and toads) with lightning (and rain) and with the lightning deity. Page 31a of the Madrid Codex, dating to around the time of the Spanish Invasion and perhaps even earlier,[42] depicts Chaak, the rain deity, surrounded by frogs that are glyphically labeled according to the four world directions with water all around (Thompson 1970:258). Present-day Tzotzil Mayan speakers from Chenalhó believe the lightning-rain deity "lives in the interior of a mountain, the doorway to his home being a cave guarded by a frog" (Thompson 1970:268). It is said that "the caves in the hills surrounding the Tzeltal community of Pinola are inhabited by many powerful deities—Thunderbolt, Meteor, Whirlwind, and the Sombrerón ... These animal [spirit] companions are kept in the caves and fed special food by the toad, who is the wife of Thunderbolt. Thunderbolt is the chief of everything" (Vogt and Stuart 2005:172). Similarly, the wife of the Chol Mayan lightning deity is a toad (Cruz Guzman et al. 1980:116). For the Huastec Mayans as well the wife of the chief lightning-rain deity is a wrinkled old toad; and frogs are known to the Huastecs for announcing the coming of rain (Alcorn 1984:140). The lightning deities also dance and drink in their caves with their female frog consorts according to Taube, who identifies the chief lightning deity (Huastec Muxi' Maam) as God N of the Maya codices (1992:97).

The Yucatec rain ceremony named *ch'a chaac* (bring rain) requires four frog impersonators, boys who are tied to each of the four posts of the altar, and who croak like frogs while the shaman prays (Redfield and Villa Rojas 1932:142) (**Figure** 45; see color plate section). Girard supplies the information that Chortis have a rain ceremony, similar to that of the Yucatecs, in which four youths personifying the frogs crouch under the altar table and croak like frogs (1995:137). Atop the table and in the center is a basin (or canoe) filled with water into which are placed a pair of frogs and a pair of fish, the basin symbolizing the earth's navel. "The Chorti elders believe the frogs and fish they place in the canoe are sacred entities and they call them 'Angels,' a term they

---

[42] Michael Coe, however, "believes the Madrid Codex postdates 1624.... He maintains that the paper with Latin text is sandwiched between two layers of bark paper, evidence that it was incorporated into the Codex during its manufacture" (Schuşter 1999:26).

also apply to the rain gods. They affirm that 'they know better how to ask for the benedictions of grace'" (Girard 1995:136).

Joyce states that at the Classic period Zapotec city Monte Alban, "by the Nisa phase, zoomorphic figurines in the form of frogs are common and perhaps related to rituals involving water and rain" (2009:142). That Mixtecs today relate toads to lightning can be easily inferred from the following passage: "When lightning struck a house in one of the Nuyoo hamlets, people speculated that a toad might have gotten inside and crawled under a bed, since lightning, because of its strong link to Rain, seeks to destroy the children of Tachi" (Monaghan 1995:139). Snakes are the most salient children of Tachi, but clearly toads are also his children. This explanation of a lightning strike is similar to the logic used by Chorti Mayans with respect to lightning axes being sent to destroy Chicchans.

Images of the frog-faced Aztec water goddess Chalchiuhtlicue, she of the Jade Skirt, "sometimes represent her as a frog, a creature sacred to and specially protected by Tlaloc" (Phillips and Jones 2005:196), and recent excavations at the Templo Mayor in Mexico City reveal a large number of objects related to the lightning deity, Tlaloc, including an altar with stone frogs (Broda 1987:214, 216). This of course recalls the contemporary Chortí altar that during the rain ceremony has a centrally located basin of frogs.

To sum up, ample evidence from various groups in Mesoamerica links frogs (and toads) to the lightning deity, and therefore to lightning. Their fondness for water and the fact that they announce the rains might be enough to explain, if not predict, their relationship with the rain deity.[43]

Lightning and Fish

As mentioned above, rains of fish and of frogs are rather commonly observed phenomena (McAtee 1917:221), far more frequently reported in newspapers than in scholarly articles. Rains of fish are even more common than of frogs, apparently, and there is at least one well-known rain of fish that has been occurring every year for more than fifty years in the department of Yoro in Honduras (cf. Kallen 1991:6). Known as the "rain of fishes" (*lluvia de peces*), this phenomenon has generated many newspaper accounts that usually describe it as happening between May and July, beginning with dark clouds, thunder and lightning, and heavy rain for a couple of hours, after which many small fish are found on the ground, gathered up by local people and brought home for dinner (cf. Chandler and Prado 2007: 110). This constitutes a clear connection between

[43] Well-known Aztec metaphors for the earth are the toad and the crocodile. There is also evidence that some Mayans share and may earlier have shared in these metaphors. The rain deity's name in parts of Guatemala, "Mountain Valley," meaning "earth (lord)," helps to explain the consistency of these metaphors.

fish and lightning as manifested in rainstorms. It is spectacular, relatively rare, and thus memorable.

Taube connects the lightning deity to fish and fishing by means of iconographic analysis, relying on images of Chaak from late Preclassic times as well as much later images of God B in the Dresden Codex. He also mentions a contemporary Guaquitepec Tzeltal belief that the San Anhel (lightning deity) is both god of lightning and god of fishing, apparently related to the observation that after lightning strikes water, dead and stunned fish remain for the gathering (1992a:23–24). The latter phenomenon is also well known in other parts of the world (see e.g., Fonvielle 1871:144). Recalling the basin with a pair of fish and a pair of frogs that sits atop the altar table during the Chortí rain ceremony, it is significant that the elders see them a sacred and call them "angels," which is one of the terms for the rain deities. Fish too, then, have an association in Mesoamerica with lightning and the rain deity, if a rather less frequently encountered association than that of frogs.

Lightning and Red

Shells of the *Spondylus* (thorny oyster, spiny oyster)[44] can range in color from blood red to orange to pink, and even to cream color. The deeper red colors in particular were most appreciated by the Aztecs and the Maya of Classic times. It is likely that this appreciation was related to the mollusk's color suggesting blood, the association with water, its natural habitat, and the difficulty in obtaining these shells, which when living are found many meters beneath the ocean's surface (Miller and Taube 1993:153). It has been maintained too that the spiny shells that are sometimes the color of blood suggested blood sacrifice, which in Mesoamerica was and still is associated with fertility (Paulsen 1974:44).

Classic Maya images sometimes depict the rain deity with a serpentlike zoomorphic face, serpents often dripping from his mouth, and wearing a *Spondylus* shell ear decoration. As Bassie-Sweet notes, "In hieroglyphic writing, the name Chahk is most often represented by a portrait of this deity. One of the hieroglyphic writing conventions is to reduce a logographic sign to a single element. This element is then used to represent the entire sign. In the case of the name 'Chahk', the *Spondylus* shell earring is often used to represent his entire portrait" (2008:7). This surely constitutes a strong association between lightning and the *Spondylus* that represents the deified form of lightning in the symbolism of Maya imagery. The red color of the more desirable *Spondylus*

[44] Among the species of *Spondylus* found in coastal American waters are *S. princeps*, *S. ictericus*, and *S. americanus*.

shells is an important association here. We need not ignore the other attributes of the *Spondylus* either; its watery home and its reminder of blood sacrifice due to the spines and the color, as well as the fertility augured by such sacrifice. These too show a linkage to lightning.

The Maya maize deity often has a *Spondylus* shell at his waist, and Maya women are frequently pictured with a waist level *Spondylus* shell covering their genitals. This is said perhaps to be related to associations of fertility (Miller and Taube 1993:153), and though we cannot be certain, it would be fitting if that notion of fertility were connected in Maya minds to the monthly menstrual blood that is part and parcel of human fertility.

Other kinds of evidence also associate lightning with the color red.[45] For example,

> the portrayal of thunderbolts as small children or dwarfs is found across highland Guatemala. Saki C'oxol is described... as a dwarf who moves like lightning... [and] characterized as the keeper of animals and the spirit of volcanic mountains... dressed in red.... In contemporary Momostenango the C'oxol is a red, dwarf thunderbolt god who appears in creation stories... struck the first K'iche' elders (father-mothers) with his ax and awakened the sheet lightning in their blood. (Bassie-Sweet 2008:105)

In a discussion of which of the thunderbolt deities was able to split open the rock mountain to get at the maize, Bassie-Sweet (2008:105) suggests the power primacy of red lightning over green lightning in Poqomchi' folklore, noting that Youngest Thunderbolt succeeded in blasting open the stone mountain in which maize was hidden prior to human access when Sudden Thunderbolt (*raxa* "green/blue") had been unable to accomplish this task. Bachajón Tzeltal speakers say that of the three lightning bolts that tried to split open the rock mountain in which maize was hidden, it was Red Lightning Bolt that finally succeeded (Slocum 1965). Pitarch (1998) explains that Cancuc Tzeltal Mayans view green lightning (*yaxal chauk*) as weak while red lightning (*tzajal chauk*) is strong. Spero (1987) confirms the power dominance of red over green lightning among Chol Mayans. The Yucatec Maya rain-lightning deity known as X Thup Chahk is the smallest and most important of the rain-lightning gods, most commonly associated with the eastern direction, which is associated with the color red (Redfield and Villa Rojas 1934:114–115, 137). So the more powerful of the lightning deities is the red one.

---

[45] Reichel-Dolmatoff notes a similarity between helpers of the Mesoamerican rain-lightning deity and the Master of Animals in Colombia. "The Mexican *chaneques* are also the supernatural masters of game animals and fish, and here a new parallel with Colombian Indian cultures arises. Among the Tukano tribes the Master of Animals is imagined as a red dwarf who dwells in caves or at the bottom of deep pools; he is closely associated with the jaguar" (1972:59).

The name of the Yucatec rain god Chaak is nearly homophonous with the word for "red" in more than one Mayan language. In Yucatec /cháak/ means "rain" and is the name of the rain-lightning deity, while /chak/ means red (Bricker 1998).[46] In Tzotzil Mayan *chauk* is "lightning" and *chak* means "red." In Kekchi Mayan (*raq'*) *kaaq* is "lightning" and *kaq* means "red." Even some four millennia ago, proto-Mayan **kaq* "red" and proto-Mayan **kahoq* "lightning, thunder" (Kaufman and Norman 1984:117) were composed of similar sounds. While evidence from sound similarities of words with different meanings by itself does not persuade a linguist that the words share conceptual features, it is at least suggestive that for some individuals there may indeed be a conceptual linkage that could ultimately be explained by either etymological relationships or by psychological associations attending the similarities, especially when the referents appear to share some features on the basis of other evidence, as "lightning" and "red" do.[47]

Chevalier and Sanchez Bain (2003:164) remark that the powers of redness, thunder, and lightning are used by the Sierra Popoluca to counter health threats from heat, suggesting that the reason for this is located in requirements of strength, which arises from heat, without which the body grows cold and ultimately dead. The juxtaposition of red and lightning in this context too suggests a strong association.

Tenochtitlán, capital city of the Aztecs, had two shrines atop the pyramid known as the Templo Mayor. One side, dedicated to Huitzilopochtli, was blue; the other, dedicated to Tlaloc, was red. Bancroft, in a careful description of the sculpted and painted Tlaloc deity in the chapel in the great Temple of Tenochtitlan, speaks of Tlaloc's having a green feather headdress and front teeth painted red (1875:325). A more recently discovered pot depicting Tlaloc's face shows red teeth, red earflares, and a large red crown. Some say that Tlaloc was dressed in red with a green feather headdress, but codical pictures tend to show him dressed in blue with red highlighted spots, and with large green-feathered headdresses.[48]

Imagery of Teotihuacan's lightning deity, given the Nahuatl name Tlaloc because of his iconographic resemblance to the Aztec lightning deity Tlaloc, donated imagery and associated concepts to Classic Maya iconography of war and of bloodletting:

---

46 The rain-making ceremony mentioned above is /ch'a'ah cháak/ and refers to "seizing the rain" (Bricker 1998).

47 It is worth noting that this near homophony in the Mayan language family regarding "red" and "lightning" is not found in other Mesoamerican language families.

48 A large painted surface above a cave in Oxtotitlan, Guerrero, Mexico, depicts an Olmec figure wearing an "owl" mask with Tlaloc-like "goggle eyes" surmounted by a green headdress and wearing various green ornaments, suggests by its cave entrance location and a couple of Tlaloc-like features that this could be a rain-lightning deity. He is seated on what has been identified as a jaguar mask (Grove 1970:frontispiece).

> The non-warfare oriented association of the Maya Tlaloc becomes evident when his effigy appears as a deity presiding over bloodletting and other ritualized events in Maya artwork. Here, bloodletting is symbolic. The blood, which is extracted and sacrificed, is a metaphor for rain. Tlaloc, the god of rain and lightning, is present when these events occur. It is possible that bloodletting is a symbolic way of embodying the spirit of Tlaloc to help bring nourishment. (Groff 2003:20)

It is not surprising that the red color of blood would be associated with lightning through the Tlaloc deity's imagery, borrowed by Classic Mayas from highland Mexican Teotihuacan, especially since obsidian, viewed by many as a product of lightning, was used for ritual bloodletting in Mesoamerica, as well as in implements of war.

Recalling the red sprite of the Quiché (Quiaka C'oxol) and other examples of red beings associated with lightning, a technical paper by geophysicists is of some interest, titled "Intense Continuing Currents following Positive Cloud-to-Ground Lightning associated with Red Sprites," and defining red sprites as red-colored structures in the mesosphere that appear above thunderclouds following positive cloud-to-ground lightning discharges (Bell et al. 1998). The paper is technical and makes no mention of folklore or beliefs. Its topic and conclusions suggest the possibility that Mesoamericans, among others, have also noticed this phenomenon, accounting at least in part for the linkage in their belief systems of red with lightning.

Between 1903 and 1907, a meteorologist observed the visible colors of lightning. He saw forked lightning in fifty-seven storms and sheet lightning on seventy-eight occasions. Although several colors were noted, in sheet lightning, red, yellow, and white alone were each seen on nine occasions, other colors on fewer occasions, and green was not seen alone (Moore 1910:312–313). Another observer says that the unaided eye finds the light of forked lightning to be usually white, whereas sheet lightning is mostly red (cf. Schellen and Huggins 1872:422, 442, 608). Currently received opinion is that most of the colors attributed to lightning are due primarily not to the properties of the lightning itself, but to the attributes of local atmospheric conditions through which the lightning is perceived. Red and reddish-orange sprites are an exception to this, however. Regardless of the cause, of course, what is important for our purposes is what the peoples of Mesoamerica may have been able to view with naked-eye observation, what they may have concluded from it, and how they integrated that information into their worldview.

Mesoamericans, then, relate the color red to lightning and to lightning spirits or lightning deities in several ways. It is of at least minor interest that contemporary scientific observers also find that some lightning, when carefully observed, has associations with the color red.

Lightning in Mesoamerica: Summary

An examination of lightning's Mesoamerican associations reveals a complex of related ideas generated largely from human observation of the natural world. These symbolic associations have been integrated together in different communities in different and quite specific ways to create relatively coherent and largely consistent worldviews in which lightning always plays an important role. Abstracting from specific information gleaned from various sample communities, it is possible to describe a more general Mesoamerican cosmology involving lightning and rain, along with related weather phenomena. The description necessarily involves indigenous perceptions of their detrimental as well as their beneficial effects on the rest of nature; their symbolic as well as observed associations with people, animals, plants, and geography; and the cognized causal relations among the components of the native cosmology.

Lightning and attendant rains when seen by Western eyes appear to be directly delivered from the skies, generally springing from the clouds, but in indigenous Mesoamerica they are thought to be sent by the lightning lord (variously known as the rain god, mountain lord, or Earth Owner), a deity best approached by way of caves seen as connecting this world with an Other World. The Other World is inhabited by ancestors, deities, spirits, souls, various monstrous beings, and some extraordinary forces, as well as by some plants and animals familiar in this world. Caves, then, constitute the ultimate source of lightning and rainstorms and are naturally an important site for human petitioning to the deities responsible for the weather. Petition is required because the deities can reward or punish humans. Lightning can snuff out a life in an instant, and the milpa can be destroyed by too much or too little rain. When gratified by human appeals and sacrifices, the lightning deity can nurture and stimulate crop growth and provide potable water for people. Observation makes it clear that some individuals are more successful than others in their appeals to the lightning lord, and this success has been over the years often attributed by design or default to individuals with success in hunting, or war, or especially in mediating cures for others in the community, often seen as the most important ability attributed to a shaman.

Whether as an agent of the lightning deity or harnessed to human bidding, lightning and rain represent powerful forces operating for benefit or detriment in the lives of individuals and the fortunes of communities in Mesoamerica. To Western minds this smacks of notions involving good and evil, and so indeed, after five hundred years of occupation of their land by foreigners with Catholic and other Western religious-inspired worldviews, some indigenous Mesoamericans also see these forces as being those of good and of evil, instead of simply desirable or detrimental, or simply powerful or not.

Some "shamans," and perhaps other individuals, are perceived to be able to control these forces for good or for evil (or for desired or feared outcomes), at least to some extent. Being struck by lightning and surviving is an overt indicator of one's connection to lightning and control over it, for example, and for that reason it is not surprising that survivors of lightning strikes are often seen as thus embarking on the shamanic path. This constitutes one form of initiation for the shaman. It could be seen also as the imparting of important curing or other knowledge from lightning to the individual, as inferred from and related to the fact that some lightning shamans get diagnostic information from what they report as "lightning" running up or down their legs. The dream, trance, and hallucinogen-inspired worlds provide other avenues for shamanic initiation. Shamans are believed to have the power to communicate with or travel to the Other World in order to cure disease, bring rain, and protect their communities. By virtue of this power they are often thought also to be able to cause illness, promote bad weather, and attack other communities. Shamans and other powerful people are frequently associated with such potent forces as lightning, or such dominant and potentially deadly animals as serpents, jaguars, and eagles, top predators in their respective worlds of air, earth, and water (or underworld).

The relations of both dependence and interdependence that indigenous Mesoamericans have had with other entities in the world around them, when combined with observations of coincidences and correlations that could be attributed to causal relationships, might easily have spawned an explanatory worldview in which each individual has a co-essence or "animal familiar" whose destiny is to some extent shared with that individual—an entity that might in appropriate circumstances serve to assist the individual and that in some situations might even participate in a transformation from one into the other. As mentioned above, the notions of shared essences and therefore shared destinies between humans and certain other animate things (e.g., animals, lightning), and of certain humans having some control over these other things, can be found in diverse communities in the world and are likely to have arisen independently multiple times and in different places.

Given such a worldview, it is appropriate that shamans and other powerful people will be the ones to have co-essences that are powerful, dominant, and deadly in their respective spheres of influence, co-essences such as lightning, jaguars, eagles, and serpents. Animal familiars, and access to the Other World for those with powerful familiars, are understandably part of a worldview that can explain much of what can be observed.[49]

---

[49] For example, if a healthy person dies quite suddenly, it might be explained as witchcraft, it can be explained as caused by another's familiar, or it might have been his own familiar suffering some mishap. These multiple possible explanations create a system that can filter out many possible contradictions.

Two things suggest themselves as models allowing for the genesis of such animal familiars as are part of the Mesoamerican worldview. First there is the apparent "psychic connection" that many have observed and or felt between twins, particularly identical twins, and sometimes between other relatives, or occationally even friends. Second, various persons have noticed a similar phenomenon between certain individuals and their domesticated animals and pets, something especially noticeable with dogs, but that also occurs with certain people and particular wild animals.[50] With such models people can envision and presume similar supposed connections between individual humans and animals either restricted to such cases as those which provided the models, or perhaps broadened to include such animal-human relationships for all people. Broadening the requirements to include certain weather phenomena as well as various animals as co-essences is but a small step, and one more easily attributable to Mesoamerican worldviews, in which such weather phenomena are animate.

The effects of lightning can be seen as either rewarding or punishing individuals and communities, and to the extent that lightning is seen as animate or wielded by an animate anthropomorphic deity, or both, humans can make requests of it within a logical framework based on perceived human needs, desires, and goals, with the expectation that the lightning deities will respond in accord with their own needs, desires, and goals, presumably similar to those of humans. Veneration might take the form of sacrificing to the gods such human food items as tortillas or small animals, or offering special food, such as chocolate, delightful-smelling incense, or the heated wax and smoke of burning candles. Prayers or chants singing the praises of the appropriate deities might be offered as well. Sometimes only a specialist will know what the deities desire or will be able to employ magical means based on iconic, indexical, or symbolic relationships.

To adequately address all the offerings entailed by multiple occasions of veneration by many persons, the lightning deity has miniature and often multiple counterparts that aid in accomplishing their responses to the articulated desires of humans engaging in appropriate veneration. The forms of veneration are often based on a model derived from family interactions and other social relationships that are also informed by family and are likely to involve notions of power, reciprocity, persuasion, and sacrifice.

---

[50] Current Western recognition of the phenomenon of such psychic unions and unspoken communication (perhaps something like sharing a soul) has come in the form of film and video titles such as "horse whisperer," "dog whisperer," and "bear whisperer."

Being humans, and with limited imaginations, we must invent our gods to some extent in our own image.[51] These gods are then likely to have humanlike needs for sustenance, a home, assistants, and so on, although given that they are more in control than humans, their needs would likely not be identical to those of humans. Mesoamericans invented an anthropomorphic deity to be in charge of thunder, lightning, and rain, and they venerate him in hopes of channeling his power to their benefit, beseeching him to bring rain when needed and not to bring it when it would harm the milpa. Hoping for neither deluge nor drought, they venerate a deity that can take different forms, that is both powerful and benevolent, yet he can be destructive as well if his ceremonial and respect requirements are not met. The destructive side of this deity's nature is able to protect favored individuals and communities by harming their enemies. His benevolent aspect provides those he favors with plenty while withdrawing it from their enemies.

This lightning deity maintains a domicile and various shrines so that he can be approached with requests and gifts. In Mesoamerica the lightning deity's house is a cave on the mountainside, and atop the mountain there is not infrequently a shrine resembling in at least some respects a human house. The mountain is where wild plants grow and wild animals go; it is undomesticated space, a necessary storehouse for plants and animals that are hunted and gathered. The mountain is also a storehouse for water, ultimately derived from rain, originating in its purest form from springs on mountainsides that feed the creeks and rivers farther down. The mountain is where lightning can be seen to strike more frequently. One approaches the lightning deity for veneration in places that are special, places that might reach down into the bowels of the earth where there must be another world. Caves do this, as do springs, and lakes.

Because in addition to relatives, people in Mesoamerica have animal familiars or co-essences, it seems appropriate for lightning to also have animal familiars. The animal familiars of lightning are powerful, related to water (the material element perceived as associated with and produced by lightning), and to earth, where lightning lives, and to the sky, where lightning is most visible. Thus certain powerful animals of each of these domains have been recruited as animal familiars for the lightning deity.

Since humans as leaders working for the community have helpers, it makes sense that lightning would also have helpers doing the work of making rain and lightning, as well as recruiting lightning shamans. One might expect the helpers, being less important than the deities themselves, to be smaller, and

[51] Guthrie (1993) has developed a theory of religion that, among other things, plausibly explains the evolutionary advantages of perception templates that result in anthropomorphic (and zoomorphic) deities.

indeed they are.[52] That is perhaps why dwarfs symbolize the underworld and essential chthonic power, doing the bidding of their master, whether the bringing of rain for crops or hail as punishment for insufficient adherence to traditional ceremonial practices including veneration of the gods. Like children, dwarfs are miniatures, sharing certain qualities with them. Both can be seen as assistants and also as essences. They can go places where larger beings cannot, and regular human dwarfs may well have symbolized to the Mesoamericans of old, some of the extraordinary powers of such beings as the "rain dwarfs" that constituted part of their cognized environment and worldview; beings that could slip easily into small openings in the rocks where humans could not follow, and beings that might appear or disappear in a flash.

Lightning is accompanied, as with a twin, by thunder, so it is not surprising that lightning and/or thunder are associated with twins in many parts of the world. Nonetheless there are few indications in Mesoamerica that lightning is seen directly as related to twins. The dyad of thunder and lightning is so integrally related to and accompanied by rain in Mesoamerica that the three can be said to form a triad of sorts, although this particular triad is little recognized as such in Mesoamerica. Nonetheless, lightning in several ways is presented and described in terms of threes, and it has various triadic properties in Mesoamerica, particularly in the Mayan region.

Mushrooms, serpents, frogs, and several other principal components of Mesoamerican worldviews with connections to lightning and rainstorms are additional elements from which community cosmologies were constructed—cosmologies that have been, of course, continually evolving and differentiating within and between local ethnolinguistic groupings in Mesoamerica. These associations have mutated through time, as with any narrative, through misunderstandings, reinterpretations, narrator's license, and so on, and in this process they have been transformed into multiple and diverse forms that are to some extent self-correcting or generative of reconverging notions due to their frequent basis in empirical observations. Thus despite continuing internal diversification of traditions, the diffusion of ideas from one community to another through trade and other contacts has resulted in widespread Mesoamerican sharing of significant elements that have been isolated, documented, and generalized here.

The sharing of various lightning-related components of Mesoamerican cosmological systems provides further impetus for treating this as a relatively

[52] This fits with young children being smaller than their parents and is in accord with the frequently encountered iconographic rule of powerful leaders being represented as larger than their followers. It is notable too that dwellers inside the mountain/earth are generally thought to be smaller than beings living on it, just as dwellers in the sky will be thought of as larger. This appears to be so in most parts of the world where information is available on the subject.

unified culture area. Moreover, the evidence that has been provided illustrates how associations that at first may seem incongruous actually have some logical and empirical consistency. One can see, for example, how warfare can be associated with nourishment, bloodletting with fertility, and witchcraft with curing. And it also becomes clear how through the various observable and cognized characteristics of lightning, it can be thought to reward (e.g., through nitrogen fixation, creating shamans, bringing rain) and to punish (e.g., frightening, maiming, or killing individuals and/or their crops).

4

Discussion

LIGHTNING IN THE ANDES AND MESOAMERICA

This comparative study of lightning in the Andes and Mesoamerica was carried out in part to determine whether similarities and differences in the patterns and associations brought out herein would address larger questions regarding distinctions and similarities in pre-Columbian religious thought, legend, and folklore. It became apparent early in our collaborative research that direct comparative analysis would pose problems in presenting contextualized generalizations, so we have deliberately produced separate presentations of lightning in local worldviews as it relates to these culture areas. Our rationale was to provide a basis for differentiating the various contexts in which it occurs. Our reluctance to produce a trait list or checklist of traits present or absent in each region was precisely because that would have entailed separating "traits" from their contexts, potentially misrepresenting their meanings. Our respective summaries of the data have been marked by necessity from our individual understandings of meaning, relevance, and coherence of the available evidence. In addition, comparing it piecemeal would violate both our individual understandings and obscure information that might be gleaned from our individual filters to clarify the meanings of our interpretations. Within this framework of abstracting from separate accounts of the two regions under consideration there are elements, taken from each account that can be compared and contrasted. These elements can be placed back into their original contexts by returning to our separate accounts.

Our presentation of these data is also related in part to the comparative homogeneity of language in the Andes as opposed to the heterogeneity in Mesoamerica. The meanings and associations of lightning are diverse and often specialized in Mesoamerica, with more than seventy languages in that region, represented by speakers in many more separate communities than that. Even if one ignores the specific individual details in the ways in which lightning fits (and fitted through time) into each of their many different worldviews, one must nonetheless consider at least three broad regions relatable to four different language families (Oto-Manguean and Mixe-Zoquean, focused on Oaxaca, Mexico; Uto-Aztecan, focused on the central Mexican highlands; and Mayan in Yucatan, Chiapas, Guatemala, and Belize). In the central and southern Andean region little remains to us other than the various Quechua dialects and Aymara language families. Linguists have demonstrated a high degree of convergence reflecting an intense contact between Quechua and Aymara speakers from the earliest stages of their language dispersals, and as a consequence they have a great deal of vocabulary in common as well as a host of striking structural parallels. These structural convergences reflect clear and deep associations between these language families. Such linguistic similarities are clearly apparent in the language surrounding lightning as well as its religious and symbolic significance, its cultural classification, and related meteorological associations. These similarities appear to reflect remarkably similar concepts, folklore, legends, and perceptions pertaining to lightning over a vast geographic area. Such unification was based in part on economic interdependence between coastal and highland cultures and the vast road networks that connected different subregions of the Andes to one another. These data support previous assertions by various prehistorians that the Andes became increasingly interconnected by networks of roads well before the rise of the Inca Empire.

It is clear that the Inca and their predecessors effectively created a linguistically unified culture area long before the arrival of the Spaniards. The linguistic data also have important implications for Inca expansion and the political unification of what were essentially linguistically and culturally diverse subject populations with regard to veneration of Illapa, as well as the cult of the sun and moon. The Inca based their authority in part on mythological grounds by extending themselves through their origin mythology to previous creation cycles and intermittent periods of destruction and chaos. These symbolic associations were based in part on a cosmogony that was pan-Andean and thus readily comprehensible to most subject populations. The cultural unification of the Andes is best expressed in the archaeological record by the existence of cultural horizons. This supports recent linguistic research pointing to a causal relationship between language divergence and the appearance of widespread and artistically similar material correlates in the archaeological record (Heggarty 2008; Heggarty and Beresford-Jones 2010). Indeed, despite numerous attempts by archaeologists to identify cultural horizons of the scale and geographic scope that characterizes the Andes, they have not been identified in the same scale. The evidence suggests that the linguistic and cultural diversity of Mesoamerica may explain, at least in part, why similarly large and expansive cultural horizons have not been documented archaeologically in Middle America (see, e.g., Rice 1993). Archaeological evidence from the Andes supports previous assertions that widespread integration of an artistic style or cultural horizon is associated with the spread of religious cults, and this is supported by various researchers with regard to lightning, which to varying degrees reflects a commonly held religious ideology and worldview (Willey 1962, 1999; Burger 1988, 1992). In the case of the Andes, such a worldview of a sacred landscape would explain in part the widespread integration of lightning related beliefs and rituals as well as their associated technologies and artistic expressions throughout this region seemingly throughout pre-Columbian times. These interdisciplinary lines of evidence at first sight suggest a large number of features and associations to lightning that are similar between these two culture areas.

Similarities

Lightning veneration appears to be ancient in both regions, apparently extending back in various forms and cultural contexts at least to the formative periods. Celestial bodies and their associated cycles as well as meteorological phenomena such as lightning were usually anthropomorphized in Mesoamerica and in the post-Conquest Andes, and lightning is associated with "mountain lords" in both regions. "Mountain lord" is referred to in Quechua as *wamani* or *apu*, and sometimes as "father" or "grandfather," and was an indirect reference to founding ancestors. Interestingly, the Mixe-Zoquean languages of Mesoamerica use the term *apu* in reference to "grandfather," while Mayan languages use the word for "grandfather" as an epithet for the lightning deity or mountain lord. Lightning in these culture areas is associated not just with rain, but also with water (lakes, rivers, ponds, lagoons, etc.) and in particular to places near subterranean springs. In the Andes, caves are associated with some lightning *huacas* or *ushnus* and with the remains of *Illapa huacacuna* or "sons of lightning," thus venerated ancestors buried there. In Mesoamerica, caves are the abode of the lightning deity and strongly associated with lightning through this symbolic connection. In both regions, the association of lightning with water, and bodies of water, has metaphorical and symbolic significance to accessing the Other World or, in the case of the Andes, the *mundo interior* or underworld. Such connections are clear with regard to lightning and the coming agricultural cycle as a bringer of rain as well as destruction of crops through hail, flood, and torrential rain.

Lightning is also associated with pre-Columbian rulership and in recent times with shamanism. Some pre-Columbian rulers adopted lightning as co-essences in Mesoamerica, while Inca rulers such as Pachacuti Yupanqui adopted the lightning bolt as his *wawqi* or *wayqe* (brother or alter ego). Other Inca rulers had fictive kin relations with *huacas* and oracles—another form of alter ego.

Contemporary shamans in both culture areas are frequently "recruited" by surviving lightning strikes. Such shamans sometimes also have familiars that are lightning, or feel "lightning" in their bodies as part of the divinatory process. In the Andes, some shamans are actually thought to be transformed into lightning to conduct their rituals and perform divination. In both regions fire plays an important part in shamanic rituals and in relation to lightning. In Mesoamerica meteors and fireballs are viewed as a kind of lightning. In the Andes shamans are recruited to the ideology of consumption associated with ecstasy and ecstatic trance, in that the transformation of the spiritual realm becomes visible through fire, which can be induced by lightning. The consumption of mythic beings is often perceived as a cosmic conflagration.

Felines also have a relationship to lightning and to shamans in both regions. This is particularly the case with the *tigre*, or jaguar (*Panthera onca* L.), because its roar is said to resemble or mimic thunder. Jaguars represented the co-essence or familiar par excellence for pre-Columbian rulers throughout these culture areas, and such associations appear to continue in various ways with present-day shamans and *curanderos*. The jaguar has a natural association with water and is one of the few species of cat that swims, thus providing a connection to the underworld based on the logic of observation.

In Mesoamerica, serpents are currently the most important symbol and familiar of the lightning deity, and this is also the case for some regions of the Andes, where the association is indirect through rainbows and dark cloud constellations in the Milky Way. Rainbows are perceived as giant serpents that rise out of subterranean springs, pass through the sky, and then bury one of their two heads in a distant spring. They are sometimes depicted in iconography and on pottery as serpents breathing lightning into one another. Thus rainbows are a manifestation, in reptilian form, of the forces of procreation and fecundity within the earth. For example, dark constellations in the center of the Milky Way appear during important points in the rainy season (October through April, especially December through February) and, when visible in the night sky, are referred to as "the serpent." There appears to be associations among dark constellations, serpents, and water in the Andes. Similarly, meteorological serpents (rainbows/*amarus*) appear only during the rainy part of the year, and they exhibit a seasonal activity cycle similar to that of terrestrial reptiles (Roca 1966). The symbolic associations between lightning and serpents in the Andes may also be related to the electric eel (*Electrophorus electricus* L.), the only species of the genus, and, moreover, exclusive to South America. Thus it would appear that some associations, particularly those associated mythologies surrounding emergence into a present creation cycle from bodies of water, lagoons, ponds, and so on, and lightning bolts have had their basis in observational and experiential logic. On the other hand, lightning (Illapa) was an animating essence of powerful symbolic and metaphorical reference. Lightning in many Andean cultures has associations to the creator, Viracocha, as well as to creation or world cycles, to mythological origins, and, with regard to the landscape and terrestrial space, to intermediate cycles or periods of destruction or chaos (Sullivan 1985). In both culture areas the natural world is perceived as an expression of an intrinsic relationship between matter and spirit. Ancient religious beliefs involving the veneration of the forces of nature, or deities with such associations, clearly indicate that the celestial realm and world below both play prominent roles in their association with such deities and natural forces as thunder and lightning.

Lightning has close associations to sacred places through bolts that descend from the sky and strike bodies of water. These associations thus link the celestial realm to the underworld through lightning, as so clearly expressed through indigenous concepts such as Pachatira and Pachatierra (Casaverde Rojas 1970:150; Nuñez del Prado 1974:246–247). Such radically different levels of existence (light and sound) are often felt to interpenetrate and coexist.

As in Mesoamerica, some Andeans perceive the underworld as populated by little people (as well as by local protective spirits and, after the Conquest, by the "devil"). Such beliefs continue in the present in various folktales and legends. In Mesoamerica, these little people are in the discourse of almost every indigenous language. They are often thought to be minions of the mountain lord (lightning deity). Similar associations occur in native Andean culture.

Lightning bolts in both regions, as well as in several other parts of the world, are often seen as the source of stone tools, including those of obsidian, flint, chert, chalcedony, and other silica-based stones. In the Andes this relation extends to transparent or translucent quartz (and perhaps other) crystals. In Mesoamerica riches and good fortune are associated with lightning via the lightning deity. In Quechua regions of the Andes, riches and good fortune are linguistically related to the root for the current term for lightning. In both regions, child sacrifice was practiced in connection with lightning, albeit with different religious rationales, and rain and hail are closely associated particularly for both their beneficial and destructive effects on crops.

Lightning in both regions is to varying degrees associated with the fertility and fecundity of the earth and the agricultural cycle. In Mesoamerica frogs and toads are also associated with lightning and the lightning deity because of their associations with thunderstorms. Certain species of toads and frogs appear quite suddenly in association with thunderstorms, thus constituting a symbolic connection with the weather, and thus with lightning and lightning portents. In the Andes, amphibian fauna in some cases have a similar association with lightning, through the weather, the annual cycle, and for purposes of divination. The Andean toad species *Bufo spinulosus* is thought to foretell bad fortune when they appear in great numbers after thunderstorms at the onset of the rainy season. Their mating and croaking behavior is often observed for purposes of divination particularly with involved with predicting the onset and intensity of the coming rains (Roca 1966:58–59). Therefore, toads are associated with weather and indirectly with lightning and divination. Toads and amphibians in both areas are cosmologically and culturally significant in various ways.

In both the Andes and in Mesoamerica the *Spondylus* shell was viewed as precious, and in both regions there was an association with lightning and attendant rain as well as with females and sexuality. The natural range of the *Spondylus/Strombus* dyad extends from Baja California to southern Ecuador and northernmost Peru (Keen 1971). Therefore, all the *Spondylus* and *Strombus* shells associated with pre-Hispanic sites south of these areas of the Andes had to be brought in the context of long-distance exchange (Paulsen 1974; Hocquenghem 1993a, 1993b). *Spondylus* and *Strombus* shell are depicted in ancient iconography and effigy vessels throughout the cultural sequence after the Initial period in Peru, and Late Formative in regions to the north, suggesting that such symbolic associations and religious importance have very ancient origins in these regions, since such interaction first appears in the Formative periods and continues until the Conquest. The red-colored exterior rims of such shells were apparently the most precious part of these shells in both of these culture areas, perhaps because the color symbolizes blood, while the spines may have suggested the bloodletting in Mesoamerica and female sex with reference to the fertility of the earth in the Andes. In Mesoamerica there was the additional association of the color red with the lightning deity as well as with lightning, and the fact that these shells were found deep in the ocean could have contributed to perceptions that they represented the watery underworld. Undoubtedly, difficulties in obtaining a constant supply of these species of marine shells contributed to their appreciation among elites as well as commoner populations in both regions.

In the Andes *Strombus* conch shells were commonly provided as ritual offering to lightning *huacas* and ritual festivals related to veneration of Illapa and of a whole host of deities and high-status individuals, as prominently depicted in art and iconography throughout the pre-Columbian sequence (Paulsen 1974; Sharon 1978, 2001; Donnan 2004; Cordy-Collins 2001b). In Mesoamerica, *Strombus* and other large gastropod genera (now referred to as conch shells), some with highly suggestive pink coloration, served as wind instruments, but iconographic evidence for a direct association with lightning or the lightning deity is not readily apparent.

Lightning in both culture areas is symbolically linked to fire, not surprisingly, since lightning represents primordial heat and can start fires. The consumption of mythic beings is often perceived as a cosmic inferno, because it transforms the spiritual and symbolic world. In many pre-Columbian societies, humans were differentiated from all other life by our ability to control fire. In both regions, there are various lines of indirect evidence suggesting that it was mastery over fire that provided a basis for communicating with celestial realms of primordial heat (Sullivan 1988). Mythological accounts in both regions mention that it was the possession of fire that differentiated humans, from other animals, plants, and spiritual entities. This is apparent when noting that the transformation of spirits or animating essences becomes visible by fire. This may explain why chroniclers claimed that the Inca Huascar put the principal lightning *huaca*, Catequil, to fire. Mastery over fire is seen as intrinsically linked to the ability to directly communicate with celestial realms of primordial heat, and thus to lightning, as well as to ideologies of consumption associated with ecstasy and ecstatic trance as in the case of shamans or *curanderos*. Shamans are considered in both regions as masters of fire and the symbolism of fire centers the world, explaining in part why the three-stone hearth was so important in many regions of Mesoamerica as symbolic of the sacred center (Carrasco 2010:603). In both regions, individuals struck directly by lightning bolts and who survived were considered to be particularly powerful. In both regions, they are often perceived as having access to the spiritual realms and sometimes given special status within their society. Shamanism has its origins in part in divination, the perceived ability to anticipate or predict future events, and to interpret the past using "true knowledge" of the spiritual realms (Reichel-Dolmatoff 1972; Sharon 1978, 2000; Sullivan 1988). Given the importance of domesticated plants to the ancient diet in both regions, such divinations often include abilities or responsibilities involving weather control, predicting the coming of the rains, and so on. Thus we may infer that such knowledge involved accessing the spiritual and cosmic realms, which to varying degrees had direct or indirect reference to lightning in both regions. In summary, it would be fair to say that there are numerous similarities through the various cultural perceptions surrounding lightning in the Andes and in Mesoamerica, some of which are quite striking, as are the differences and what these imply regarding distinctions between these various regions.

Differences

While there are indeed similarities in the various associations of lightning within the cultures of both regions, there are also numerous differences, some clearly due to the regional environmental differences, others to language and culture. The relative linguistic homogeneity of the Andes as opposed to the diversity of Mesoamerican languages is revealed in the widespread similarities in the material culture, as well as rites, ritual, and religious ideologies associated with lightning in the former region and clear distinctions among the various cultures in the latter. The differences are most readily apparent in festivals and rituals associated with the agricultural and cosmological calendars of these regions, which to some extent is related to geographic and climate differences. For example, most thunderstorms in the northern and central Andes occur between December and February, and they are most common along the coast and particularly in the high sierra. It was in precisely these times in the annual cycle that festivals, rites, and ritual celebrations in veneration of lightning were held. It is in the context of such rituals and rites to lightning that we begin to have a basis for understanding why lightning has associations with the coming agricultural cycles and to the fertility and fecundity of the earth. On the other hand, most thunderstorms in Mesoamerica occur between May and September or October, and they are frequently accompanied by deluges of rain. Rituals and festivals in honor of deities associated with lightning were often celebrated with the coming of the rains or during the end of the cycle, when offerings, prayers, and other ritual forms of veneration are made to ward off destructive hail and flooding storms.

In the Andes, lightning also had reference to rainbows, the moon, lunar eclipses, and particular constellations, while evidence in Mesoamerica for such reference is effectively lacking. In Mesoamerica lightning is largely associated in symbolic form and through legend, mythology, and ceremonial activity with particular deities who control and dispense it, while in the Andes lightning was and still is perceived as a force of nature or an animating essence. The widespread associations of lightning to specific anthropomorphic deities and mythological beings in Mesoamerica, as opposed to cultural perceptions of lightning as a force of nature or animating essence, may to varying degrees explain other differences between the regions with respect to lightning associations. Evidence linking lightning in Mesoamerica to the moon, lunar eclipses, or dark constellations is lacking. Consequently, symbolic associations to lightning and its role in the annual calendar can be viewed as culturally conditioned. While there is compelling evidence to suggest an association between lightning and the metallurgical arts in the Andes, such associations are not apparent in Mesoamerica. However, some metallurgical technologies in western Mexico are similar to technologies in some regions of the Andes.

There are other differences too, including an emphasis on gender in the Andes that is not so pronounced in Mesoamerica. For example, in traditional Andean culture the lightning bolt or *rayo* (one that strikes the ground) is thought of as female, while cloud-to-cloud lightning (*relámpago*) is male. Andean legend and folklore perceive female bolts or *rayos* as potentially entering the womb of pregnant women who are outside during thunderstorms. When such spiritual "contact" results in male twins, they are perceived as sacred and extraordinary, essentially *huacas* to their *ayllus*, because they represent a transcendence of a gender duality: male/female. The organization of culture and the natural world in terms of such oppositional dualities is central to traditional Andean worldview. Since the Conquest, Andeans who converted to Christianity to varying degrees have come to perceive such spiritual "contact" between female bolts, pregnant women, male twin births as infants born without original sin, essentially an immaculate conception. The Order of Santiago promoted the Immaculate Conception of Mary in part because such special status was not given to female twins in the Andes. It is possible that similar perceptions provided a religious rationale for referring to male twins as "sons of lightning," or *Illapa huacacuna*, in pre-Columbian times. In Mesoamerica, lightning bolts are not affiliated with gender (though the anthropomorphic deities and their assistants are gendered), and there is no evidence of a belief that lightning bolts impregnate women or enter their wombs. During the Colonial period, such gendered symbolic associations in the Andes were readily incorporated in the cult of Santiago, the patron saint of Spain. Lightning associations to Santiago, while peripheral to many Spaniards and mestizos, made this force of nature a formidable power and source of veneration to indigenous populations in the Andes.

The apotheosis of Santiago Matamoros as a lordly, omnipotent epiphany has no equal in the post-Conquest era. Santiago was usually addressed with respect as "patron," "apostle," "señor," or, most often, by the single appellation "Santiago Illapa" in the Andes. The importance of this association is also implied by historic events on both sides of the Atlantic. The early sixteenth-century Spanish Crown commanded the knights of the Order of Santiago to defend religious belief in the Immaculate Conception of Mary. Gender distinctions between bolts and cloud-to-cloud lightning are not apparent in the Mesoamerican literature. There is reason to suppose that this may again be related to its association to particular deities, legends, and myths, while the Catequil cult of Illapa was widespread and persistent throughout the Andes. Such differences are also apparent from the fact that most indigenous Mesoamerican cultures transferred their veneration from the pre-Columbian deities to Catholic saints, particularly through conversion or the fusing of indigenous perceptions to Catholic orthodoxy early on in the Colonial periods. The syncretism of lightning by contemporary Andeans involves fusing the ancient concept with Catholic saints St. James and (to a lesser extent) Santa Barbara. On the other hand, San Juan, the twin brother of the apostle St. James (Santiago), is more frequently linked to the Mesoamerican versions of the lightning deity. In Morelos, Nahuatl speakers syncretize St. John the Baptist with Tlaloc the Aztec rain god, while those in the Puebla highlands call him "Santiago" (Ingham 1986:184). San Miguel and San Bartholomew are also associated with lightning by various Mesoamerican societies (Thompson 1970; Govers 2006). While such symbolic reference to Catholic saints is prevalent over a wider geographic area in the Andes, the cultural perception of lightning as an animating essence and part of the natural world has to varying degrees reinforced earlier legends and folklore surrounding lightning among some indigenous populations to a much greater degree.

Andean symbolic reference and religious veneration to lightning essentially represents a veneration of the natural world, as evident when boulders or rocks are cleaved by lightning bolts and the place becomes sacred or a *huaca*, often also demarcated by platforms and artificial terraces. Culturally transformed by direct lightning strikes, lightning *huacas* reaffirm the cultural significance of lightning and lightning bolts as *enqa* or *sami*, an animating essence, rather than a particular deity. Lightning in Mesoamerica was closely tied to the landscape and especially to caves through which communication with the

Other World could be achieved in connection with particular deities and venerated ancestors. This is not the case in the Andean region, where caves are not given the same importance with respect to either lightning or communication with the Other World. Evidence to indicate that places in Mesoamerica where lightning struck were demarcated and/or venerated is complicated by geographic and ecological differences, as large areas of the Mesoamerican lowlands are covered in dense jungle vegetation. However, in both lowlands and highlands there are cultural beliefs and folklore maintaining that lightning bolts striking the earth leave behind stone tools, including farming implements and weapons; flint blades, or worked obsidian, as well as jade axes. In the Andes stone axes appear with the spread of metallurgy in the Andes, and axe-monies formed a kind of currency with regard to tribute and long-distance exchange (Lechtman 1976, 1984, 1988; Hosler et al. 1990).

Another regional difference is with respect to little beings or dwarfs. In Mesoamerica the lightning deity's assistants are dwarfs or little people, usually, though not exclusively, found in association with mountains, and frequently caves. To be sure, such little people are present in Andean consciousness, but the evidence to suggest that such beings are related directly to lightning is ambiguous except perhaps through an indirect association. Along the southern Ecuadorian coast, dwarfs from the underworld or *mundo interior* are sometimes accused of impregnating women while their husbands are away at work. This parallels some other Andean folklore through cultural perceptions that women out in a lightning storm can become impregnated by a bolt that "enters" the womb, resulting in twin births or offspring with certain disfigurements (Krappe 1938; Gade 1983).

Andean religious veneration with respect to lightning strikes include cultural perceptions of a sacred geography in which the forces of nature and celestial bodies had reference to particular places in the landscape, and in the case of lightning to a person or a place transformed by a bolt. Preliminary research at the lightning *huaca* San Catequilla de Pichincha, north of Quito, Ecuador, has suggested that places presumably struck by lightning and demarcated by circular enclosures or platforms of cut stone or fieldstone may have actually functioned architecturally in relation to the associated platforms to refine astronomical calculations when using sight lines to mark celestial cycles (Staller 2011).

Sling stones appear to have a symbolic association to lightning in the Andes. Lightning *huacas* often had piles of sling stones left as offerings. Such symbolic connections are also characteristic of the Andean region through lightning's association with death, as slings were commonly used in hunting terrestrial mammals and birds, and also in the context of regional conflict and warfare. Sling stones in the Andes are usually in the shape and size of a golf ball, or consist of river-rolled cobbles which approximate this shape. Thus the shape of sling stones closely approximates hail, providing another basis for their connections to lightning and its associated rituals and rites. Their significance to Andean culture is further reflected in Inca mythology, where they are projected by one of the mythological ancestors into each of the cardinal directions, resulting in the transformation and creation of the Valley of Cuzco. There is some reason to believe that slings and sling stones have celestial association with the constellation of Orion.[1] Various chronicles state that lightning was symbolized by a constellation outlining a man wielding a club in his left hand and a sling in his right. The Inca perceived Illapa as dressed in shining garments that project flashes or bolts of lightning when he whirled his sling to bring on the coming rain. Thus sling stones are associated with lightning through Andean mythology and astronomy and are given ritually as offerings. One of the original mythological Inca hurled sling stones in four directions from the summit of a sacred mountain near Cuzco, and they struck so hard the surrounding hills were transformed and ravines were created (Betanzos 1996 [1557]:14). Therefore, the Valley of Cuzco was perceived as mythologically shaped and transformed by sling stones. Such perceptions and associations provide a basis for understanding why such stones were closely associated with Illapa in Inca religion. Obsidian and particularly quartz crystals also have symbolic associations to lightning in the Andes as ritual offerings at *huacas* and to lightning shamans in contemporary culture. Such stones are sometimes used for divination; however, they are not generally perceived as the result of a lightning strike.[2] Rather, their association to lightning in the Andes is related to the sources of such stone with mountains and mountain summits. The primary natural element that associated lightning to the underworld was water, particularly underground springs, lagoons, ponds, lakes, and streams, rather than caves (Topic et al. 2002).

In the Inca capitol of Cuzco, a faceless gold statue, kept in a special sanctuary, represented the religious idol to Illapa or lightning. Moreover, this association, while specific to Inca culture, was extended and then incorporated into much of the empire, where it was symbolically represented by lightning oracles or had reference to particular lightning *huacas*. Non-Inca communities or *llactas* were literally connected to such *huacas* and to the Inca state through Capac Hucha sacrificial rites and to the venerated ancestors buried in such places. In Andean culture, lightning was always linked to the landscape and ancestor veneration and not as closely to particular deities. Mesoamerican lightning deities were related to the surrounding landscape by their symbolic connections to the cardinal directions or the intercardinal quadrants and their

---

[1] Note that the Classic and Colonial period lightning triad of deities identified by Karen Bassie and by Allen Christenson is associated with stars in Orion as well as with the three stones of the hearth, creation myths, and maize.

[2] Quartz crystals are important tools of divination used by shamans in Mesoamerica and North America as well, though in Mesoamerica, at least, they appear to have little connection to lightning.

associated colors. Sometimes such deities or their assistants were differentiated according to forms of water or moisture (i.e., hail, rain, clouds, and fog). Such differences speak directly to distinctions in religious thought regarding symbolic and metaphorical associations of deities to meteorological phenomena and sacred places in the geography and how such connections were perceived. In the case of the Andes, these were in some cases literal associations to a particular place, through architectural modifications and/or to burials of venerated ancestors, Capac Hucha victims, and "sons of lightning," or *Illapa huacacuna*. In Mesoamerica, such connections were through what were essentially anthropomorphic or zoomorphic deities and related to locally significant places such as specific mountains and caves because these locations provided portals for communication with the Other World. In contemporary cultures, these portals are said to be traversed by deities, spirits, and ancestors.

Although both regions illustrate a symbolic and mythological association between lightning and twins, in Mesoamerica the relationship between lightning and twins is indirect, inferred from the relationship of both to serpents. The association of twins and lightning found in the Andes, on the other hand, is both explicit and pervasive. The fact that various other regions of the world, including to the north of Mesoamerica, link twins to weather phenomena and especially to lightning is particularly interesting in view of the lack of evidence for a direct association of this sort in Mesoamerica. Andean male twins or individuals born with harelips or cleft palates were "sons of lightning." This is related to the perception that any thing or person referred to as a *huaca* had some extraordinary characteristic or quality. Twin births or facial abnormalities were often seen as an empirical indications of having been touched by this force of nature before birth, thus they were considered *huacas*, sacred or extraordinary. In the Andes, when individuals with such abnormalities or twins died, they would be embalmed and put into ceramic urns kept and/or buried at sacred places that had been struck by bolts of lightning. These "sons of lightning" were ascribed with a special status, and children (both male and female) sacrificed in Capac Hucha rites were given sacrificial offerings and revered as *huacas* or founding ancestors by their communities or *llactas*—hence the direct and symbolic association with mountain lords or *apu* (Salomon and Urioste 1991 [c. 1598–1608]; Sharon 2001). Significantly, it is in this association to mountain summits that the Andes and Mesoamerica share lightning traditions.

In the Andes lightning is directly linked to the principles of rebirth and renewal and the coming rainy season because lightning was believed to control and have power over various meteorological phenomena. Andean cosmology links the heavens to the terrestrial realm; thus lightning is also associated with solar cycles and eclipses of the moon, whereas these associations are not so easily recognized in Mesoamerica. Such differences may in part reflect a fundamental difference in the pre-Columbian religions of these culture areas. Lightning was a primary spiritual entity in the Andes as opposed to an aspect, attribute, or instrument of a primary deity such as Chaak or Tlaloc, as is the case in Mesoamerica. Some of these differences in these cultural areas are therefore cultural: an emphasis on a pantheon of deities versus direct veneration of forces of nature pertaining to the terrestrial (earthly) or the celestial realms. Lightning is seen first and foremost in the Andes as a force of nature with both deadly and life-giving power. These differences are also in varying degrees related to the extreme geography of the Andes and the fact that many *huacas* and *ushnus* (regardless of their symbolic associations) were also used for astronomical purposes and thus closely tied to the agricultural cycles—knowing when to plant and harvest food crops.

In the Andes, lightning is known to transmit the decisions of major spirits, but we have little evidence that lightning transmits such decisions in Mesoamerica. Another distinction regarding such beliefs in the Andes is that lightning *huacas* or *ushnus* could make other such *huacas* "speak," and in some cases lightning *ushnus* were seen as "related" to one another through fictive kinship ties. Such concepts and relationships are absent in Mesoamerica. These differences are no doubt tied to Andean organization of the geography and communities (*llactas*), into binary dualistic, triadic, and quadripartite divisions (Gelles 1995; Zuidema 2011). Such classifications and divisions of society and the geography cannot be conflated in Incaic thought. Fictive kin relationships and terminology indicated relationships between or among categories, while the Inca structured modalities as complementary and not interchangeable (Staller 2006:449, 453). Andean cultures are primarily organized as moieties, their social organization markedly dualistic, and thus lightning is perceived as a transcendence of sound and light and, because they are intrinsically interrelated, as an apprehension of categories. Lightning is usually preceded or followed by thunder, thus it is an indirect reference to the underworld. Andean worldview perceives this world and the Other World (underworld) as coexisting in the same experience, only in an ambiguous and negative way.

Sounds and the visual aspect of lightning in Mesoamerica may refer to animating forces, and lightning, like all things perceived as alive, has an animating power or force. It is unclear in Mesoamerica how lightning specifically contributes to, or whether it simply reflects, animating essences or life forces. Lightning is recognized as having great power, however, and humans who survive a lightning strike are often seen as having been chosen for shamanic initiation. Mesoamerican societies appear to make a consistent linguistic distinction between the visual and auditory aspects of lightning, while usually relating both sight and sound to the same deities. Converted Andeans often rationalize surviving a lightning flash as due to the mercy of God or one of his saints, particularly Santiago Illapa, and in some instances as a shamanic initiation. Andeans make linguistic distinctions between thunder and lightning, but couch the associations in the folklore surrounding lightning and twin births.

In Mesoamerica the mountain lord is on occasion perceived in dualistic terms paralleling the thunder (sound) and lightning (sight) dichotomy. For example, among the Kekchi Maya a thunder deity, Mam, is thought of as old and malevolent, while Tzultacah, a lightning deity, is characterized as younger and more benevolent. A different age-related dichotomy has been reported among Huastecs (Alcorn 1984). Such young and old dualities are found in other Mesoamerican deities, as, for example, with the maize god (see Taube 1996a, 1996b), and may reflect the importance of relative age. Relative age in Mesoamerica is important to the extent that older and younger siblings are distinguished terminologically in the same way that gender is. Another duality in Mesoamerica relates to animal familiars: serpents tend to be more closely associated with the visual effect, while the jaguar is more closely associated with thunder, the auditory effect.

Lightning plays a central role in the transcendence between life and death in the Andes, particularly when individuals survive a direct strike. Such occurrences are seen as an indication of shamanic initiation, and the perception is that such individuals can access the spiritual realms, particularly the underworld (Gade 1983). While lightning in Mesoamerica is known to play a role in life and death by occasionally causing the latter, it is not a central role. Lightning and its associated deity are often linked to shamanism and trance, particularly to the power of rulers. The death deity is entirely separate from the lightning deity. In the Andes associations of lightning are manifested in the form of fictive kin relationships with sacred places in the surrounding landscape.

Rainbows are associated with rain and therefore also linked to lightning, and in the Andes they are seen as harbingers of hail and lightning and are believed to presage rainstorms.[3] The available data from Mesoamerica indicates that they are seen as holding back rain or signaling the end of a thunderstorm, having no clear-cut connection to lightning or the lightning deity.[4] When rainbows in the Andes move across the sky they are thought of as willful; however, most are malevolent. Legends about how they usually steal from men are common, but when associated with "female" lightning bolts, they can

---

[3] Such perceptions are particularly apparent in the high sierra, as rainbows can be seen for some distance while storms may still be approaching the viewer. Rainbows are generally seen as a sign that the storm has subsided, so such perceptions are to some extent related to the extreme Andean geography.

[4] Rainbows in Mesoamerica are generally viewed as dangerous, entering and exiting earth through caves, sinkholes, and sometimes anthills (Ingham 1986:116; Sandstrom 1991:249), and causing illness such as headaches, frights, or soul loss. Pointing at a rainbow can cause the pointing finger to blacken, wither away, or become hopelessly bent. Some believe that a snake is to be found under a rainbow, making it dangerous to pass by (Guiteras-Holmes 1961:235). If present at all, any connections of rainbows to lightning are subtle.

move into the womb, causing great stomach pain, and in some cases even inducing miscarriages. While rainbows are often described as malevolent in Mesoamerica, the perception that lightning causes stomach pain, twin births, or physical deformations is not apparent in the literature. Since lightning has strong reference to rainbows in the Andes, certain colors have reference to lightning in the form of ritual offerings or in association with certain rites of passage. In the Andes, red, black, and white serpents and maize have importance to rituals and curing surrounding lightning veneration. This may be because the colors black and white refer to the underworld and celestial realm, respectively. Black, white, and red popcorn varieties are used to make *chicha* consumed and provided as offerings in the context of rituals associated with *Illapa*. There seems to be no particular association of lightning veneration with the colors black and white in Mesoamerica; rather, the primary lightning reference is with lightning deities that in the Mayan region relates them to red and green, and less directly to colors associated with the cardinal directions.

In the Andes the sun, thunder, and lightning have three manifestations. In the Mayan region during the Classic and Colonial periods, lightning also was given a triadic classification, with three different names, but the sun was not related to these names. In the Andes there is an astronomical association of dark cloud constellations and Orion with lightning veneration. There seems to be no evidence linking lightning to dark cloud constellations in Mesoamerica, nor to any other constellations, apart from its putative connection with the "belt" portion of what is known to us as the Orion constellation. Lightning is strongly associated with the mountain lord, Earth Owner, or lightning deity among various Mesoamerican cultures. The association of lightning with celestial constellations in the Andes is a result of Incaic beliefs that Illapa controls thunder and, by extension, all celestial bodies and climatic phenomena, particularly rain, hail, and rainbows. If the Inca are any indication, it is also related to the fact that Andean cultures appear to incorporate the celestial realm into their cosmologies. Such symbolic associations are related to the power of lightning to connect humans, primarily rulers and shamans, to the tripartite cosmos—that is, the celestial, terrestrial, and underworld realms.

The most dramatic expressions of lightning are associated with life crises and rites of passage in the Andes. This is related in part to the perceived effects that lightning has on humans and its role in associating non-Inca people to the state. In Mesoamerica evidence is lacking concerning associations of lightning directly with life crises and rites of passage, though such associations may have existed in the past. Lightning conveys the power and beneficence of all nature, venerated to provide fertility to coming agricultural cycles. The Mesoamerican lightning deity was perceived as powerful, and was primarily petitioned to provide agricultural fertility through his perceived power over meteorological phenomena associated with rainstorms. In the Andes, Illapa was venerated as a force of nature, not a deity.

The importance of lightning *huacas* and *ushnus* to channeling fluids and water ideology is evidenced by the underground channels or canals and drains found with many of the Andean truncated platforms, as well as various carved rocks, fountains, and basins that channel fluids across their carved surfaces at such sites. Such patterns of association are related to the idea that the channeling of fluids is symbolic of the male generating principle as it relates to the landscape and agricultural fertility. In Andean worldview, the earth is conceived as female, as Pachamama or Earth Mother, and the fluids channeled into agricultural fields bring forth agricultural fertility, thus the cultural connection to the male generating principle. In the Andes it appears that all religious veneration to lightning as well as to the sun and moon has reference to places in the landscape or the natural world—places that have either been directly affected by such forces of nature or that are linked to individuals, communities, and other sacred places in the landscape. In this extreme environment irrigation technology and the control of water linked communities to one another and to the state. In pre-Columbian Mesoamerica some evidence suggests that on sacrificial occasions blood and likely alcoholic beverages were channeled into holes in the ground as offerings to the lightning deity (earth lord). Similar Mesoamerican channeling of blood and alcohol into the ground has been recognized in the colonial and contemporary literature (e.g., among the Chorti), but a connection with lightning has not been established.

5

Conclusions

We have discussed lightning in this book as a phenomenon of nature that can inspire terror as well as wonder. Sometimes fearsome, with awesome power and striking beauty, lightning is likely to be incorporated in one way or another into the worldviews of all or nearly all the communities of the world and, in many, to be perceived as having both creative (fertility and energy) as well as destructive (instant killer and fire starter) powers affecting the lives of humans in multiple ways. We know that lightning can smash homes, set fires, and kill people; but it can also provide some of the nitrogen necessary for plant growth, and can bring rain to slake the thirsts of crops as well as people. Lightning is usually accompanied by clouds that obscure the sun, yet it is bright enough even in the darkness of gathering rain clouds or in the blackness of night for moments at a time to rival the light of the sun. The occurrence of lightning alternates with, or in some cases is accompanied by, other weather events, such as strong winds, hail, fog, and of course the clouds from which it usually springs—elements that could be thought of as coordinate phenomena, independent of one another, possibly causally related to one another, or perhaps related mainly by virtue of being under the control of some animate being, such as a deity.

Although different societies doubtless have their own ranges of internal variation with respect to classifying, talking, and thinking about lightning and related weather phenomena, the fact that most humans share similar sensory and perceptual apparatus leads us to conclude that the range of possibilities from one society to another, and in fact the ranges of both intra- as well as intercultural variation, are likely to be relatively narrow. Thus we expected to find both similarities and differences between Andean and Mesoamerican worldviews regarding lightning, but not major differences, and we expected the sorts of similarities present to involve lightning associations that might be produced naturally within societies on the basis of observable features of lightning and the kinds of inferences that one might make about those features with minimal interference from culturally loaded information.

Some other similarities might be expected, of course, on the basis of elements of shared cultural ecology; specifically, at the time of the Spanish invasions both regions had large populations, sophisticated agriculture, with some of the populations living in cities. Moreover much of their populations were governed, if in some cases only loosely, by states that were partial empires. Finding similarities in lightning-related worldviews between these distant regions of the Americas that could not be relatively easily explained by, or at least attributable to, observational logic and/or similar features of cultural ecology would imply contact between the regions. Differences between the regions regarding lightning-related worldviews would be expected to fall into two categories: those related to differences in ecology and therefore to how lightning would fit into the ecological situation, and those not so related, which could therefore be attributed to cultural differences between the regional cultures. The Spaniards colonized the New World and occupied both regions for roughly the same half a millennium. They brought with them a language and religion that had a profound effect on indigenous populations in these regions. There are clear distinctions regarding the associations of lightning to different saints, particularly the twin brother of St. James, St. John the apostle. These differences no doubt relate to the colonial history of different regions and subregions and may also be calendric, relating to when the feast day of a particular saint (e.g., Santa Barbara or John the Baptist) falls within the agricultural cycle. Chroniclers emphasize that most ritual celebrations or rites of passage associated with lightning were held at the beginning of the planting or rainy season.

Our comparative analysis of lightning in Mesoamerica and the Andes has uncovered some important similarities and differences, which speak directly to fundamental distinctions in pre-Columbian religious thought and social organization. Perhaps the most notable pattern uncovered in the Andes is the degree to which all religious veneration has direct reference to the natural world rather than to particular intermediary deities or attributes of deities, as in Mesoamerica. Although caves and water (subterranean springs, lagoons, lakes, etc.) had significance to lightning in the Andes, it was usually a symbolic reference to accessing the tripartite cosmos usually through animal familiars. Animal familiars in the Andes were associated with the annual cycle and agricultural fertility, and often with maize consumed in various forms or left as offerings, particularly in the form of *chicha*. Access to different realms of the tripartite cosmos may be reflected in both regions by the repeated triadic classifications with regard to lightning and its associated manifestations. In the Andes, however, access to the spiritual realm was more literal and direct, while in the case of the underworld usually indirect or metaphorical.

To the extent that humans could pretend to bargain successfully for their own destiny in the face of lightning's dual character (helping and harming), they might well have needed to create intermediaries to intercede and/or be dealt with more successfully than humans might directly with whatever forces create the lightning, given its awesome power. Individuals struck by lightning who could remain alive would be obvious candidates for being such intermediaries, just as would the lightning deity's own assistants if this deity were an anthropomorphic being. In Mesoamerica, shamans and the deities and their helpers are such intermediaries, as once were the rulers of the various polities. In the Andes shamans, animal familiars, and the *huacas* and *ushnus* were intermediaries of this sort prior to the Conquest.

In the Andean region the rulers were also such intermediaries. Andean rulers would "communicate" with oracles, and some lightning *huacas* could communicate with one another or "make other *huacas* speak." Such *huacas* and *ushnus* were often used for astronomical purposes or as ceremonial centers, which provided communities with a cultural and spiritual understanding and veneration of the surrounding landscapes. Mountain peaks and places struck by lightning were marked by the spiritual presence of such meteorological phenomena. Serpents, jaguars, and amphibians had such symbolic associations because of their literal connections to water or rainstorms, or, in the case of immaculate conceptions, by direct or perceived contact with *rayos* or bolts, as evidenced by certain deformities or by twin births. Communities (*llactas*) and powerful elites were literally connected to places associated with Illapa through fictive kinship or human sacrifice and burials of founding or venerated ancestors placed on mountain summits, platforms, or caves at lightning *huacas* or *ushnus*. Moreover, recognition through veneration of such places and the *Illapa huacacuna* buried or entombed there related to aspects of social organization and traditional forms of reciprocity between the state and non-Inca subjects, ultimately representing recognition of the state's authority. The widespread similarities in expression of lightning veneration in the Andes is suggested to be related to the linguistic homogeneity of this region with regard to Quechua and Aymara and the spread of the lightning cult in associations with cultural horizons. It is most apparent in the ethnohistoric accounts and with the expansion of the Inca state because the Spanish Crown sought to replace such veneration with the cult to Santiago.

In contrast to such associations and referents, literal, symbolic, and metaphorical, in Mesoamerica lightning is associated with specific pre-Columbian deities, rulership, and shamanism, as well as with the concept of co-essences or animal familiars. In both regions, shamans are transformed into lightning, have lightning familiars, or feel it in their bodies in the context of ecstatic trance and divination. Rather than fictive kin relationships or "communication" with oracles as well as between oracles, Mesoamerican communication with the spiritual realms was through sacrificial offerings, prayer, and the use of fire in places associated with the lightning deity. In both regions, lightning was associated with agriculture and the cardinal directions, but in distinct ways. Mesoamerican rain and lightning deities were thought of as multiples of four and were linked to the four directions and their associated colors, and as forms of moisture (e.g., hail, rain, clouds, and fog) that could be wielded by the lightning deity. Such associations and distinctions appear to extend back to pre-Columbian times and beyond. The power of lightning and its connection to mountains and summits is evident in mythology when lightning splits the Mesoamerican Sustenance Mountain, revealing the maize, or when the mythological Inca stands on the sacred mountain summit and transforms the Valley of Cuzco with sling stones shot in the cardinal directions.

Many of the similarities between Andean and Mesoamerican conceptions of lightning can in fact be seen as extensions of observations that could be made independently by the inhabitants of these regions, based on the nature of lightning. In both regions lightning and rain co-occur, and since rain is so important for crops, an association with fertility is to be expected. Because successful harvests are associated with riches in agricultural societies, lightning and riches are also perceived as connected. Since lightning is very powerful and can be a real danger to human life, it is not surprising that one who survives a lightning strike can be assumed to possess extraordinary powers—perhaps the powers of a shaman—and that these powers may be related to lightning and rain (control over water and its agricultural benefits, and the power to control fire, among other things). A person rendered unconscious by a lightning strike, or put into a state of mental dissociation by that means, might easily be perceived as in a trancelike state allowing for enhanced communication with entities from and in other realms of existence.

Thunderbolts are loud as well as potentially lethal, and these qualities reinforce their association with warfare, drums booming, axes and other weapons clashing, sling stones thrown and landing, and so on. That the Andean sling stone and the Mesoamerican flint blade can be instruments of warfare (as well as of hunting) leads to a natural association with lightning for both regions (thunder as well as the weapons are both loud and lethal). Serpents, with their frequent zigzag pattern of locomotion, are also associated with lightning bolts in both regions, another association very likely based in observations requiring little cultural influence. Tobacco has a more limited regional connection to lightning, despite the common observation that tobacco smoke resembles the clouds that may portend thunderstorms, and despite the fact that tobacco smoking in both northern South America and in Mesoamerica is perceived to facilitate communication with deities.

Mushrooms, associated with lightning in Mesoamerica and on occasion in the Andes, may share this connection on the basis of independent observations of the natural world: the fruiting caps of mushrooms frequently spring up immediately after a rainstorm. Similarly, frogs often appear in abundance after rainstorms, and this connection is clearly made in Mesoamerica, if not so well attested in the Andes. While the color red can be connected to lightning by naked-eye observation, it seems also to be a linkage made primarily in Mesoamerica. Fish are associated with lightning and rainstorms in Mesoamerica, and rains of fish have been observed in various places in the world, if not in the mountainous Andes, where fish are also not associated with rainstorms or lightning.

If observations of lightning's behavior can account for many, or even most, of the shared associations with lightning in both Mesoamerica and the Andes, notions of diffusion, stimulus diffusion, and/or contact are therefore not required to account for the shared similarities between the Andean region and Mesoamerica. Then if the differences in associations can be attributed to ecological differences and to the differences in historical circumstances and in worldviews that inform the traditions of both regions, the symbolic and cultural associations of lightning in the two regions being different again tell us little or nothing about whether or not there was cultural contact between the regions.

Another striking pattern with regard to differences in these regions as reflected in worldview is the degree to which Andean religions represented a worship of nature, in that most religious veneration referred to the natural world, to sacred places in the landscape, or to the celestial realm. Lightning in Mesoamerica, on the other hand, was usually associated with a deity, one in charge of thunder and rain as well as hail and the clouds, and often the wind. This deity, as the mountain lord, was also said to be the "master of animals." In other words, the object of veneration in Mesoamerica was an anthropomorphic deity with presumed power over such forces of nature and not the force of nature itself, which was the venerated object in Andes. Such deities or beings were represented symbolically with reference to regional mythologies and cultural identities with particular cultures and civilizations. They had different names from one language to another, and some variation exists with respect to stories about them. What is conspicuous about the Mesoamerican situation, however, is that despite the regional variation in language, ecology, and culture with respect to lightning ideology, there is a remarkable number of similarities, including motifs in oral narratives and coherence of relevant worldview, and this attests to a Mesoamerican culture area or convergence area, suggesting a great deal of contact between peoples in Mesoamerica. Apparently for the most part the contact did not involve deliberate attempts to create language shifts until the Spaniards arrived.

In the Andes, forces of nature such as lightning, or celestial bodies such as the sun, the moon, or Venus, were venerated themselves, without the mediation of intervening anthropomorphic deities (at least before the arrival of the Spaniards), and this may also provide a basis for understanding the widespread similarities regarding lightning. The fact that someone hit by a lightning bolt, as in the case of male twin births, was given special status, rank, and protective rights in the Andes speaks directly to the extent to which the idea of a sacred landscape predominated ancient worldviews in this region. In Mesoamerica lightning and rains are believed to have been sent by the rain god in earth's interior. This is why caves constitute the ultimate source of lightning and rainstorms in this region, while in the Andes it was water, ponds, lakes, and lagoons that provided connections to the layers of the tripartite cosmos. Since lightning is a force of nature (i.e., not controlled by deities) in the Andes, it is most commonly associated with high mountain summits and mountain passes and often symbolized by crosses. Such sites are often the resting place of venerated ancestors, as in the case of Capac Hucha burials, and perceived as *apu* or mountain lords, one of the preferred locations for their associated *huacas* and *ushnus*. Although both cases involve mountain summits and the burial sites of venerated ancestors, these associations are to varying degrees a reflection of observational logic and worldview. As this study has shown, they reflect fundamental differences between ancient pre-Columbian religions in these regions: in the Andes the religious ideology is essentially a veneration of nature, while in Mesoamerica their ideologies have much more in common with what we know about ancient religious ideologies in general, particularly their associations of deities to rulers and their political and religious authority.

The various lines of evidence presented in this analysis on lightning have required reference to other forms and kinds of religious veneration in these regions of the pre-Columbian New World. They certainly provide compelling evidence to suggest that lightning played a central role in pre-Columbian religions, as it did throughout the world.

When we began this project, we recognized the possibility that the comparisons and contrasts between the regions under study might contribute to further understanding the nature and/or extent of any contacts that might have occurred between them. There have been persistent attempts to link the Andes to Mesoamerica over the years, more recently positing direct contact between Ecuador and western Mesoamerica through Andean voyagers who visited Mesoamerica and western Mexico by sea and then returned to their point of origin. Hosler (2009), for example, drawing on post-Invasion sketches and some engineering calculations, proposes that expeditions from Ecuador went in two-masted balsa craft up the Pacific coast to the western coast of Mexico, searching for *Spondylus* shells along the way and bringing both metal samples and metallurgical knowledge to groups in Mexico. Samples from western Mexico suggest that some metal technologies new to Mesoamerica but already known in Peru and Ecuador much earlier had arrived by 650–700 CE. One can infer that successful sea voyages at that time could have been accomplished earlier, leaving evidence that has not yet been uncovered archaeologically.

Malmström (1995), starting from posited but not demonstrated linguistic evidence of suggested similarities of Purépecha (Tarascan) in Michoacán to

Quechua of Peru, seeks to position the intrusive Tarascan group speaking a linguistic isolate as descendants of a few boatloads of migrants from Peru or Ecuador around 800 CE. He concludes that the Tarascan migration and expansion in Michoacán was an event following some two thousand or more years of continuing contact by sea between the western coast of Mesoamerica and the Andean region of South America. Despite the lack of solid linguistic evidence, the scenario is nevertheless plausible. The Tarascans are an intrusive group and probably arrived somewhat before 900 CE. They were acquainted with metallurgy and utilized stone architecture reminiscent of Andean stonemasonry, while also possessing stirrup-spouted ceramic vessels, again reminiscent of some Andean ceramic ware.

Anawalt offers an interesting and relevant biological conundrum in the fact that the painted jay (*Cyanocorax dickeyi*), which has a limited western Mexican distribution and is found nowhere else in North America, has as its closest relative the white-tailed jay (*Cyanocorax mystacalis*), which is found only some 2,400 miles to the south and has a distribution that is limited to coastal Ecuador and Peru (Anawalt 1992:115). Proposing that importation by way of sea trade would constitute a feasible explanation for this anomalous distribution of the two species of the same genus, she then goes on to focus her well-documented article on clothing, persuasively arguing that

> clothing styles, design motifs, and techniques of cloth production found in codex illustrations and on pottery and extant textile fragments suggest diffusion of culture traits from the northern coast of South America to West Mexico and on into the American Southwest. The non-Mesoamerican garments depicted in a West Mexican sixteenth-century manuscript and on mortuary figurines buried more than 1,000 years earlier in an adjacent area find analogs only in styles that were present in Ecuador from 1500 B.C. up to the time of Spanish contact. Clothing and textile design motifs represented on figures found in the West Mexicans haft tombs of Ixtlán del Río, Nayarit, indicate that these parallels existed as early as 400 B.C. A variety of other data suggest that intermittent maritime contact persisted between Ecuador and West Mexico through the intervening period and into the sixteenth century. (Anawalt 1992:114)

In a book that was published in 1994 and therefore just missed the opportunity to address Anawalt's evidence, Bruhns (1994) devotes much of a chapter to a brief and cogent discussion of previous suggestions of contact between South America and Mesoamerica, skeptically examining evidence from shaft tombs to metallurgy, from balsa rafts to *Spondylus* shells, from feline motifs on materials from roughly contemporaneous Chavin and Olmec ideological representations, to ceramic forms such as shoe pots and stirrup vessels, finding problems with each as evidence of transcontinental contact. She also mentions a few plants that were dispersed from one hemisphere to the other and again raises some doubts about conclusions that others have drawn about them. One must demur with respect to her suggestion that a "multiple centers of domestication hypothesis for maize" (Bruhns 1994:364) is supported by current evidence. Surely that is no longer the case if it ever was (see Staller 2010a:122–131, 141–148, 192–196). More important than her skeptical perusal of earlier proposals about contact is her insistence that a fuller examination of all the evidence will be required to evaluate its pertinence to the more appropriate questions of not only when contact occurred, but where, between whom, to what ends, how often, and what items were exchanged, in addition to what ideas. Dispersal of material goods, ideology, and plants and animals must be investigated with more in mind than using them as evidence that contact occurred. She concludes that while there was doubtless contact between the Andes and Mesoamerica, almost all of the evidence needed to provide any sort of interesting picture has yet to be provided (Bruhns 1994).

Taking into account possible raft voyages, similarities between the regions in metallurgy, shaft tombs, ceramic form, Chavin and Olmec feline motifs, iridescent painting on pottery, copper axe-monies, use of *Spondylus*, distribution of certain food plants and other botanicals, and clothing manufacture and fashion, as well as language distribution, we concur that there was doubtless contact between the regions, but the actual nature, timing, and frequency of that contact remains to be investigated more fully. The question remains, to what extent does our comparison and contrast between the two regions regarding lightning and its associations yield information useful to supporting or denying arguments for such contact?

We conclude that our study, while useful for understanding the nature of lightning and its cultural associations in the two regions, is unable to supply telling arguments in either direction. Indeed, we feel that one of the contributions of this book lies in not only presenting distinct, but in several ways similar, lightning-related worldviews, but also in suggesting that while notions of diffusion, stimulus diffusion, and/or direct contact may indeed have occurred to produce similarities in these lightning-related worldviews, they are *not required* to account for most of the shared similarities that we have been able to adduce.

There are of course a number of remarkable similarities between these two culture areas that do not relate directly to lightning and are not easy to explain by way of similarities in natural observations or in terms of ecologies of the tropics. At first glance the specifics of associations with the number four would seem to be a case in point. In the Andes, the original mythological Inca consisted of four pairs of siblings or spouses. In Mesoamerica, the original Quiché forefathers, as described in the Popol Vuh, consisted of four pairs of spouses (Tedlock 1996; Christenson 2007). The Inca Empire was divided into four parts, and Tawantinsuyu means "land of the four corners."

Similarly, in Mesoamerica the earth was divided into four quadrants, and in many cases it was the corners rather than the cardinal points that were stressed. This is also apparent in research concerning the astronomical function of pre-Columbian architecture in these regions (Aveni 2005; Broda 2006; Magli 2005). Another similarity related to the number four in the two regions comes from the quadripartite division of land and the world. These similarities may be reflecting the importance of ancient cosmogony and cosmology, and particularly astronomy, to the reckoning of temporal cycles relating to the agricultural calendar, and, in the case of world cycles, to world creations and ancient mythologies. Ancient civilization in both culture areas kept track of solar and lunar cycles, as well as the cycles of Venus and other celestial bodies and constellations, particularly those associated with the Celestial River or Milky Way. Such celestial connections have clearly been brought out by this analysis and emphasize the importance of such esoteric knowledge in providing a basis for political authority, and the administration and redistribution of resources and labor (Sullivan 1985, 1988; Broda 2006; Zuidema 1989, 2011). Such examples and associations make it easier to see natural connections with the number four, based on astronomical observations of the path of the sun and the solstice sunrise points. Observational logic may also underlie the Andean view that the ocean (*Ticcicocha*) followed the sun after sunset on its underground journey through the deep waters of the ocean represented in some cultures by the eternal river, the Milky Way, as surrounding and supporting the earth, a view that is also present in different subregions of Mesoamerica (Broda 2006; Zuidema 2011).

We have argued, that many, and perhaps most, of the shared similarities could crop up naturally and independently regardless of whether there was some sort of contact between the regions. We explain the similarities in terms of ecological considerations with the additional assumption that humans possess similar perceptual equipment for observing the world around them and have brains that process the observations similarly in applying inductive reasoning, the notion of causation, and contextual generalization, and thus make many similar inferences from their perceptions of the world around them.

The nature, timing, and frequency of contact between the Andean region and Mesoamerica remain largely problematic and were not explored further in order to maintain the cultural and contextual integrity of lightning and its various associations to the cosmologies, folklore, and legends associated with these culture areas. To the extent that there was such long-distance contact, it will be important in each instance to ascertain whether the contact was direct or indirect; when such contacts occurred; and what sort of material goods, biological entities, and or information was exchanged, from whom and to whom, in order to document evidence of direct diffusion from one region to the other. As these interdisciplinary data indicate, lightning and its religious and cultural significance to pre-Hispanic and post-Conquest indigenous cultures in these regions was largely formed and constructed by different worldviews in which veneration in the Andes was directed to the natural world, the celestial world, and the underworld, while in Mesoamerica it was addressed to deities and mythological ancestors. All of these associations with lightning were fundamental to the survival and cultural identities of the societies to which such anthropomorphic entities, forces of nature, and places in the natural landscape were subject.

REFERENCES

Ackerman, Diane. 1988. A Reporter at Large: Crocodilians. *New Yorker*, October 10: 42–81.

Acosta, José de. 1962 [1589]. *Historia natural y moral de las Indias*. Mexico City: Fondo de la Cultura Económica.

Acosta-Solis, Misael. 1970. *Geografía y ecología de las tierras áridas del Ecuador*. Instituto Ecuatoriana de Cíencías Naturales Contribución No. 72, January 1970, Quito, Ecuador.

Adams, Karen R., and Mollie S. Toll. 2000. Tobacco Use, Ecology, and Manipulation in the Prehistoric and Historical Southwestern United States. In *Tobacco Use by Native North Americans*, edited by Joseph Winter, pp. 143–170. Norman: University of Oklahoma Press.

Alarcón, Hernando Ruiz de. 1892. Tratado de las supersticiones y costumbres gentilicas que oy viven entre los indios naturales desta Nueva España. *Anales del Museo Nacional de Mexico* 7, pp. 123–223.

Albores, Beatriz, and Johanna Broda, eds. 1997. *Graniceros: Cosmovisión y meteorología indígenas de Mesoamérica*. México, DF: Colegio Mexiquense, Universidad Nacional Autonoma.

Albornoz, Cristobál de. 1989 [c. 1584]. Instrucción para descubrir todas las guacas del Piru y sus camayos y haziendas. In *Fábulas y mitos de los Incas*, edited by H. Urbano and P. Duvoils, pp. 163–198. Crónicas de América Series No. 48. Madrid: Historia 16.

Alcorn, Janis. 1984. *Huastec (Mayan) Ethnobotany*. Austin: University of Texas Press.

Allen, Catherine J. 1986. Coca and Cultural Identity in Andean Communities. In *Coca and Cocaine Effects on People and Policy in Latin America*, edited by Deborah Pacini and Christine Franquemont, pp. 35–48. Sponsored by the Latin American Studies Program (LASP), Cornell University, April 25–26, 1985. Cultural Survival Report No. 23. Peterborough, NH: Cultural Survival.

Allen, Catherine J. 1988. *The Hold Life Has: Coca and Cultural Identity in an Andean Ayllu*. Washington, DC: Smithsonian Institution Press.

Anawalt, Patricia Rieff. 1992. Ancient Cultural Contacts between Ecuador, West Mexico, and the American Southwest: Clothing. *Latin American Antiquity* 3: 114–129.

Anderson, K. B., and W. Bray. 2006. The Amber of *El Dorado*: Class 1B Archaeological Ambers associated with *Laguna Guatavita*. *Archaeometry* 48 (4): 633–640.

Andrews, E. Wyllys, IV. 1970. *Balankanche, Throne of the Tiger Priest*. Middle American Research Institute, Publication 32. New Orleans: Tulane University.

Anonymous. 1906 [1580–1621]. Discurso de la sucesión y gubierno de los yngas. In *Juicio de limitess entre el Perú y Bolivia: Prueba peruana presentada al gobierno de la República Argentina*, vol. 8, edited by Victor M. Maúrtua, pp. 149–165. Madrid: Tipografia de los Hijos de M. G. Hernández.

Apt, Vaman Shivram. 1965. *The Practical Sanskrit Dictionary*. 4th ed. Dehli: Motilal Banarsidass.

Arías-Larreta, Abraham. 1964. *Pre-Columbian Literatures: Aztec, Incan, Maya-Quiché*. Los Angeles: New World Library.

Armillas, Pedro. 1957. *Cronología y periodificación de la historia de América precolombina*. México, DF: Escuela Nacional de Antropología e Historia, Sociedad de Alumnos.

Armillas, Pedro. 1964. Southern Mesoamerica. In *Prehistoric Man in the New World*, edited by Jesse D. Jennings and Edward Norbeck, pp. 291–329. Chicago: University of Chicago Press.

Arnold, Denise Y., and Christine Hastdorf. 2008. *Heads of State: Icons, Power, and Politics in the Ancient and Modern Andes*. Walnut Creek, CA: Left Coast Press.

Arriaga, Pablo José de. 1968 [1621]. Extirpación de la idolatría en el Pirú. In *Cronicas peruanas de interés indígena*, edited by F. Esteve Barba, pp. 191–278. Madrid: Nabu Press.

Arzáns de Orsúa y Vela, Bertolomé. 1965 [1737]. *Historia de la villa imperial de Potosí*. 3 vols. Edited by L. Hanke and G. Mendoza. Providence, RI: Brown University Press.

Assadourian, Carlos Sempat. 1989 [1568–1571]. Acerca del cambio en la naturaleza del domino sobre las Indias: La Mitá minera del virrey Toledo, documentos del 1568–1571. *Anuario de Estudios Americanos* 46. Seville.

Aulie, H. Wilbur, and Evelyn W. de Aulie. 1978. *Diccionario Ch'ol-Español, Español-Ch'ol*. Serie de Vocabularios y Diccionarios Indigenas Mariano Silva y Aceves, tomo 21. México, DF: Instituto Lingüístico de Verano.

Aveni, Anthony. 2005. *Oservadores del cielo en el México antiguo*. México, DF: Fondo de Cultural Económica.

Báez-Jorge, Félix, and Félix Darío Báez Galván. 2005. The Popoluca. In *Native Peoples of the Gulf Coast of Mexico*, edited by A. Sandstrom and E. García Valencia, pp. 139–157. Tucson: University of Arizona Press.

Bancroft, Hubert Howe. 1874. *The Native Races of the Pacific States of North America*. Vol. 1, *Wild Tribes*. New York: D. Appleton and Company.

Bancroft, Hubert Howe. 1875. *The Native Races of the Pacific States of North America*. Vol. 3, *Myths and Languages*. New York: D. Appleton and Company.

Barrera Vásquez, Alfredo. 1980. *Diccionario Maya Cordemex: Maya-Español, Español-Maya*. Merida: Ediciones Cordemex.

Bassie, Karen. 2002. Maya Creator Gods. Mesoweb: http://www.mesoweb.com/features/bassie/CreatorGods/CreatorGods.pdf, pp. 1–60.

Bassie-Sweet, Karen. 1991. *From the Mouth of the Dark Cave: Commemorative Sculpture of the Late Classic Maya*. Norman: University of Oklahoma Press.

Bassie-Sweet, Karen. 2008. *Maya Sacred Geography and the Creator Deities*. Norman: University of Oklahoma Press.

Bastian, Joseph. 1973. Qollahuaya Rituals: An Ethnographic Account of the Symbolic Relations of Man and Land in an Andean Village. PhD diss., Department of Anthropology, Cornell University.

Bastian, Joseph. 1978. *Mountain of the Condor: Metaphor and Ritual in an Andean Ayllu*. American Ethnological Society Monograph, No. 64. St. Paul: American Ethnological Society.

Bateson, Gregory. 1972. *Steps to an Ecology of the Mind*. Chicago: University of Chicago Press.

Baudez, Claude-François. 1992. The Maya Snake Dance: Ritual and Cosmology. *RES: Anthropology and Aesthetics* 21 (Spring): 37–52.

Bauer, Brian S. 1998. *The Sacred Landscape of the Inca: The Cusco Ceque System*. Austin: University of Texas Press.

Bauer, Brian S., and David S. P. Dearborn. 1995. *Astronomy and Empire in the Ancient Andes: The Cultural Origins of Inka Sky Watching*. Austin: University of Texas Press.

Beam de Azcona, Rosemary. N.d. The Legend of the Hunter and His Compadre: A Zapotec Text from San Baltazar Loxicha. Unpublished ms., University of California, Berkeley.

Bell, Timothy F., Steven C. Reising, and Umron S. Inan. 1998. Intense Continuing Currents Following Positive Cloud-to-Ground Lightning Associated with Red Sprites. *Geophysical Research Letters* 25: 1285–1288.

Bennett, Wendell C. 1946. *Excavations in the Cuenca Region, Ecuador*. Yale Publications in Anthropology No. 35. New Haven, CT: Yale University Press.

Benson, Elizabeth. 1998. The Lord, the Ruler: Jaguar Symbolism in the Americas. In *Icons of Power: Feline Symbolism in the Americas*, edited by Nicholas J. Saunders, pp. 53–76. London: Routledge.

Benson, Elizabeth. 2001. Why Sacrifice? In *Ritual Sacrifice in Ancient Peru*, edited by Elizabeth P. Benson and Anita G. Cook, pp. 1–20. Austin: University of Texas Press.

Berdan, Frances, and Patricia Rieff Anawalt. 1997. *The Essential Codex Mendoza*. Abridged ed. Berkeley: University of California Press.

Beresford-Jones, David, and Paul Heggarty. 2011. What Role for Language Prehistory in Redefining Archaeological "Culture": A Case Study on New Horizons in the Andes. In *Investigating Archaeological Cultures: Material Culture, Variability and Transmission*, edited by B. Roberts and M. Van der Linden, pp. 355–386. New York: Springer.

Bertonio, Luis. 1956 [1612]. *Vocabulario de la lengua aymara*. La Paz: Don Bosco.

Besom, Thomas 2010. Inka Sacrifice and the Mummy of Salinas Grandes. *Latin American Antiquity* 21 (4): 399–422.

Betanzos, Juan de. 1987 [1551]. *Suma y narración de los incas*. Edited by C. M. Rubio. Madrid: Atlas.

Betanzos, Juan de. 1996 [1551]. *Narrative of the Incas*. Translated and edited by R. Hamilton and D. Buchanan from the Palma de Mallorca Manucript. Austin: University of Texas Press.

Bierce, Ambrose. 2007. *The Collected Works of Ambrose Bierce*. Vol. 1. Charleston, SC: BiblioBazaar.

Billington, Ray Allen, Bert James Loewenberg, Samuel Hugh Brockunier, and David Sparks. 1962. *The Making of American Democracy: Readings and Documents*. Vol. 1, *1492–1865*. New York: Holt, Rinehart and Winston.

Binford, Lewis R. 1962. Archaeology as Anthropology. *American Antiquity* 28 (2): 217–225.

Binford, Lewis R. 1964. A Consideration of Archaeological Research Design. *American Antiquity* 29 (4): 425–441.

Binford, Lewis R. 1965. Archaeological Systematics and the Study of Culture Process. *American Antiquity* 31: 203–210.

Bolin, Inge. 1998. *Rituals of Respect: The Secret of Survival in the High Peruvian Andes*. Austin: University of Texas Press.

Bond, Donald W., Scott Steiger, Renyi Zhang, Xuexi Tie, and Richard E. Orville. 2002. The Importance of NO_x Production by Lightning in the Tropics. *Atmospheric Environment* 36: 1509–1519.

Bond, D. W., R. Zhang, X. Tie, G. Brasseur, G. Huffines, R. E. Orville, and D. J. Boccippio. 2001. NO_x Production by Lightning over the Continental United States. *Journal of Geophysical Research* 106: 27701–27709.

Bonfil Batalla, Guillermo. 1968. Los que trabajan con el tiempo: Notas etnograficas sobre los graniceros de la Sierra Nevada, Mexico. *Anales de Antropologia* 5: 99–128.

Bouysse-Cassagne, Therese. 1993. El rayo que no cesa: D'un dieu pukina a un dieu Inca. In *Religions des Andes et lengues indigenas: Equateur-Pérou-Bolivie avani et après la conquéte espagnola*. Actas du III Coloque d Etudes Andines, pp. 163–180. Université de Provence Centre Aixois de Recherches Latino-Américaines (CARLAM). Aix-en-Provence.

Brady, James E., and Keith M. Prufer. 2005. *In the Maw of the Earth Monster: Mesoamerican Ritual Cave Use*. Austin: University of Texas Press.

Breedlove, Dennis Eugene, and Robert M. Laughlin. 1993. *The Flowering of Man: A Tzotzil Botany of Zinacantán*. Washington, DC: Smithsonian Institution Press.

Bricker, Victoria R. 1981. *The Indian Christ, the Indian King: The Historical Substrate of Maya Myth and Ritual*. Austin: University of Texas Press.

Bricker, Victoria R. 1998. *A Dictionary of the Maya Language as Spoken in Hocabá, Yucatán*. Salt Lake City: University of Utah Press.

Brinton, Daniel G. 1894. *Nagualism: A Study in Native American Folk-lore and History*. Philadelphia: MacCalla and Company.

Broda, Johanna. 1987. The Provenience of the Offerings: Tribute and Cosmovision. In *The Aztec Templo Mayor: A Symposium at Dumbarton Oaks, 8 and 9 October 1983*, edited by Elizabeth Hill Boone, pp. 185–209. Washington, DC: Dumbarton Oaks Research Library and Collection.

Broda, Johanna. 2006. Zenith Observations and the Conceptualization of Geographical Latitude in Ancient Mesoamerica: A Historical Interdisciplinary Approach. In *Viewing the Sky through Past and Present Cultures*. Oxford Seventh International Conference on Archaeoastronomy, edited by T. W. Bostwick and B. Bates, pp. 183–212. Pueblo Grande Museum Anthropological Papers No. 15. Phoenix: Pueblo Grande Museum.

Brown, Cecil. 2011. The Role of Nahual in the Formation of Mesoamerica as a Linguistic Area. *Language Dynamics and Change* 1: 171–204.

Brown, Cecil H., David Beck, Grzegorz Kondrak, James K. Watters, and Søren Wichmann. 2011. Totozoquean. *International Journal of American Linguistics* 77: 323–372.

Browne, B. J., and W. R. Gaasch. 1992. Electrical Injuries and Lightning. *Emergency Clinics of North America* 10: 211–226.

Bruhns, Karen O. 1987. Los talleres de crystal de roca de Pirinchay, Provincia del Azuay. *Miscelenia Antropologica Ecuatoriana* 7 (7): 91–100.

Bruhns, Karen O. 1994. *Ancient South America*. Cambridge World Archaeology. Cambridge: Cambridge University Press.

Bruhns, Karen O., James H. Burton, and Arthur Rostaker. 1994. Archaeological Investigations at Pirincay in Southern Ecuador. *Antiquity* 64 (243): 221–233.

Bulfinch, Thomas. 1976. *Bulfinch's Mythology*. Illustrated ed. New York: Crown.

Burger, Richard L. 1988. Unity and Heterogeneity within the Chavín Horizon. In *Peruvian Prehistory*, edited by R. W. Keatinge, pp. 99–144. Cambridge: Cambridge University Press.

Burger, Richard L. 1992. *Chavín and the Origins of Andean Civilization*. London: Thames and Hudson.

Butler, H., and M. Inez. 1997. *Diccionario Zapoteco de Yatzachi el Bajo, Yatzachi el Alto, Oaxaca*. Tucson: Summer Institute of Linguistics.

Campbell, Joseph. 1949. *The Hero with a Thousand Faces*. New York: MJF Books.

Campbell, Joseph. 1974. *The Mythic Image*. Bollingen Series C. Princeton, NJ: Princeton University Press.

Campbell, Lyle, Terrence Kaufman, and Thomas Smith-Stark. 1986. Meso-America as a Linguistic Area. *Language* 62: 530–570.

Cardinale, Hyginus E. 1983. *Orders of Knighthood, Awards and the Holy See*. Buckinghamshire, UK: Van Duren.

Carleton, S. C. 1995. Cardiac Problems Associated with Electrical Injury. *Cardiology Clinics* 13: 263–266.

Carmack, Robert. 1981. *The Quiché Mayas of Utatlán*. Norman: University of Oklahoma Press.

Carmack, Robert M., Janine L. Gasco, and Gary H. Gossen. 2007. *The Legacy of Mesoamerica: History and Culture of a Native American Civilization*. 2nd ed. Upper Saddle River, NJ: Prentice Hall.

Carrasco, Michael. 2010. From Field to Hearth: An Earthly Interpretation of Maya and Other Mesoamerican Creation Myths. In *Pre-Columbian Foodways: Interdisciplinary Approaches to Food, Culture and Markets in Ancient Mesoamerica*, edited by John E. Staller and Michael D. Carrasco, pp. 601–634. New York: Springer.

Cereceda, Veronica. 1987. A proximaciones a una estéticaandina: De la belleza al tinku. In *Tresreflexciones sobre el pensamiento andino*, edited by J. Medina, pp. 133–231. La Paz: Hisbol.

Cerrón-Palomino, Rudolfo. 1987. *Lingüística Quechua*. Cuzco: Centro Bartolomé de las Casas.

Cerrón-Palomino, Rudolfo. 2000a. El origen centroandino del aimara. In *Boletín de Arqueología PUCP*, edited by Peter Kaulicke and William H. Isbell, pp. 131–142. Lima: Fondo Editorial de la PUCP.

Cerrón-Palomino, Rudolfo. 2000b. *Lingüística Aimara*. Cuzco: Centro Bartolomé de las Casas.

Cerrón Palomino, Rudolfo. 2003. *Linguistica Quechua*. 2nd ed. Cuzco: Centro Bartolomé de las Casas.

Cerrón-Palomino, Rudolfo. 2008. *Quechumara: Estructuras Paralelas del Quechua y del Aimara*. 2nd ed., revised and expanded. La Paz: Plural.

Ceruti, Constanza. 1999. *Cumbres sagrados del Noreaste Argentino: Avances en arqueología de alta montaña y etnoarqueología de santuarios de altura andina*. Buenas Aires: Eudeba.

Ceruti, Constanza. 2004. Human Bodies as Objects of Dedication at Inca Mountain Shrines (North-Western Argentina). *World Archaeology* 36 (1): 103–122.

Chandler, Gary, and Liza Prado. 2007. *Honduras and the Bay Islands*. Oakland, CA: Lonely Planet.

Chang, K. C. 1983. *Art, Myth, and Ritual: The Path to Political Authority in China*. Cambridge, MA: Harvard University Press.

Chávez, Sergio. 1992. The Conventional Rules in Pucara Pottery Technology and Iconography: Implications for Sociopolitical Developments in the Northern Lake Titicaca Basin. PhD diss., Department of Anthropology, Michigan State University, East Lansing.

Chávez, Sergio. 2006. Native Aymara and Quechua Botanical Terminologies of *Zea mays* in the Lake Titicaca and Cuzco Regions. In *Histories of Maize: Multidisciplinary Approaches to the Prehistory, Linguistics, Biogeography, Domestication, and Evolution of Maize*, edited by John E. Staller, Robert H. Tykot, and Bruce F. Benz, pp. 623–629. Walnut Creek, CA: Left Coast.

Chevalier, Jacques M., and Andrés Sánchez Bain. 2003. *The Hot and the Cold: Ills of Humans and Maize in Native Mexico*. Toronto: University of Toronto Press.

Christenson, Allen J. 2001. *Art and Society in a Highland Maya Community: The Altarpiece of Santiago Atitlan*. Austin: University of Texas Press.

Christenson, Allen. 2007. *Popol Vuh: The Sacred Book of the Maya: The Great Classic of Central American Spirituality, Translated from the Original Maya Text*. Norman: University of Oklahoma Press.

Christenson, Allen J. 2008. Places of Emergence: Sacred Mountains and Cofradía Ceremonies. In *Pre-Columbian Landscapes of Creation and Origin*, edited by John E. Staller, pp. 95–122. New York: Springer.

Christian, H. J., and J. Latham. 1998. Satellite Measurements of Global Lightning. *Quarterly Journal of Royal Meteorological Society* 124: 1771–1773.

Christian, H. J., et al. 1999. The Lightning Imaging Sensor. In *Proceedings of the 11th International Conference on Atmospheric Electricity*, pp. 746–749, Guntersville, AL.

Chuchiak, John F., IV. 2000. The Indian Inquisition and the Extirpation of Idolatry: The Process of Punishment in the *Provisorato de Indios* in the Diocese of Yucatan, 1563–1812. PhD diss., Tulane University. Ann Arbor: University Microfilms.

Cieza de León, Pedro. 1959 [1551]. *The Incas*. Translated by Harriet de Oris. Norman: University of Oklahoma Press.

Cieza de León, Pedro. 1977 [1551]. *El Señorio de los Incas*. 2nd Editorial Universo. Lima: Instituto de Estudios Peruanos.

Cieza de León, Pedro. 1984 [1553]. *Crónica de Perú: Primera Parte*. Lima: Pontificía Universidad Católica del Perú y Academía Nacional de la Historia.

Cieza de León, Pedro. 1998 [1553]. *The Discovery and Conquest of Peru: Chronicles of the New World Encounter*. Edited and translated by Alexandra Parma Cook and Noble David Cook. Durham, NC: Duke University Press.

Cieza de León, Pedro. 2005 [1553]. *The Travels of Pedro de Cieza de Léon A.D. 1532–50, Contained in the First Part of His Chronicle of Peru*. Translated and edited, with notes and an introduction, by Clements R. Markham. Elibron Classics Series. London: Adamant Media. (Originally published in 1864 by the Hakluyt Society.)

Classen, Constance. 1990. Aesthetics and Asceticism in Inca Religion. *Anthropologica* 32 (1): 101–106.

Classen, Constance. 1993. *Inca Cosmology and the Human Body*. Salt Lake City: University of Utah Press.

Cobo, Bernabé. 1964 [1653]. *Historía del Nuevo Mundo*. Biblioteca de Autores Españoles, Vol. 92. Madrid: Ediciones Atlas.

Cobo, Bernabé. 1990 [1653]. *Inca Religion and Customs*. Translated and edited by R. Hamilton. Foreword by J. H. Rowe. Austin: University of Texas Press.

Coe, Sophie D. 1994. *America's First Cuisines*. Austin: University of Texas Press.

Coggins, Clemency. 1975. Painting and Drawing Styles at Tikal: An Historical and Iconographic Reconstruction. PhD diss., Harvard University.

Coggins, Clemency. 1979. A New Order and the Role of the Calendar: Some Characteristics of the Middle Classic Period at Tikal. In *Maya Archaeology and Ethnohistory*, edited by Norman Hammond and Gordon R. Willey, pp. 38–50. Austin: University of Texas Press.

Coggins, Clemency. 1988. The Manikin Scepter: Emblem of Lineage. *Estudios de Cultura Maya* 17: 123–158.

Collier, Donald. 1946. Archaeology of Ecuador. In *Handbook of South American Indians*, vol. 2, edited by J. H. Steward, pp. 767–784. Bulletin 143, Bureau of American Ethnology. Washington, DC: Smithsonian Institution.

Collier, Donald, and John V. Murra. 1943. *Survey and Excavations in Southern Ecuador*. Field Museum of Natural History *Fieldiana* Publication 528. Chicago: Field Museum of Natural History.

Condori Mamani, Gregorio, and Asunta Quispe Huamán. 1996. *Andean Lives*. Originally edited by Ricardo Valderrama and Carmen Escalante; translated from the Quechua and with annotations and revised glossary by Paul H. Gelles and Gabriela Martínez Escobar. Introduction by Paul H. Gelles. Austin: University of Texas Press.

Conrad, Geoffrey W., and Arthur A. Demarest. 1984. *Religion and Empire: The Dynamics of Aztec and Inca Expansion*. Cambridge: Cambridge University Press.

Conrad, Geoffrey W., and Arthur A. Demarest, eds. 1992. *Ideology and Pre-Columbian Civilizations*. Santa Fe, NM: School of American Research Press.

Cook, Noble David. 1981. *Demographic Collapse, Indian Perú, 1520–1620*. Cambridge: Cambridge University Press.

Cook, Noble David. 1998. *Born to Die: Disease and New World Conquest, 1492–1650*. Cambridge: Cambridge University Press.

Cordy-Collins, Alana. 2001a. Decapitation in Cupisnique and Early Moche Societies. In *Ritual Sacrifice in Ancient Peru*, edited by Elizabeth P. Benson and Anita G. Cook, pp. 21–33. Austin: University of Texas Press.

Cordy-Collins, Alana. 2001b. Blood and the Moon Priestesses: Spondylus Shells in Moche Ceremony. In *Ritual Sacrifice in Ancient Peru*, edited by Elizabeth P. Benson and Anita G. Cook, pp. 35–53. Austin: University of Texas Press.

Corlett, William T. 1935. *The Medicine-Man of the American Indian and His Cultural Background*. Springfield, IL: Charles C. Thomas.

Costas Arguedas, José F. 1967. *Dicionario de folklore boliviano*. 2 vols., Sucre.

Covey, R. Alan. 2006. *How the Incas Built Their Heartland: State Formation and the Innovation of Imperial Strategies in the Sacred Valley*. Ann Arbor: University of Michigan Press.

Covey, R. Alan, and Christina M. Elson. 2007. Ethnicity, Demography, and Estate Management in Sixteenth Century Yucay. *Ethnohistory* 54 (2): 303–334.

Crespo, Antonio. 1983. Arquitectura de Ingapirca (Cañar, Ecuador). *Miscelánea Antropológia Ecuatoriana* 3: 194–212. Guayaquil, Ecuador.

Cruz, Pablo. 2009. Huacas olvidadas y cerros santos apuntes metodológicos sobre la cartografía sagrada en los Andes del sur de Bolívia. *Estúdios Atacameños* 38: 55–74.

Cruz Guzmán, Austencio, Kathryn Josserand, and Nicholas A. Hopkins. 1980. The Cave of Don Juan. In *Third Palenque Round Table, 1978 Part 2*, edited by Merle Greene Robertson, pp. 116–123. Austin: University of Texas Press.

De la Fuente, Julio. 1977. *Yalalag, una villa zapoteca serrana*. 2nd ed., Clásicos de la Antropología No. 2. México, DF: Instituto Nacional Indigenista.

De la Garza, Mercedes. 2002. Sacred Forces of the Mayan Universe. In *Native Religions and Cultures of Central and South America*, edited by Lawrence E. Sullivan, pp. 92–176. New York: Continuum.

Demarest, Arthur A. 1981. *Viracocha: The Nature and Antiquity of Andean High God*. Harvard University, Cambridge, MA: Peabody Museum of Archaeology and Ethnology.

Diaz del Castillo, Bernal. 1981. *The Conquest of New Spain*. 13th ed. New York: Penguin Books.

Dickerson, R. R., et al. 1987. Thunderstorms—An Important Mechanism in the Transport of Air-Pollutants. *Science* 235: 460–464.

Donnan, Christopher B. 2004. *Moche Portraits from Ancient Peru*. Austin: University of Texas Press.

Duran, Fray Diego. 1971. *Book of the Gods and Rites and the Ancient Calendar*. Translation and notes by F. Horcasitas and D. Heyden. Norman: University of Oklahoma Press.

Duvoils, Pierre. 1967 [1582]. Un inédit de Cristóbal de Albornoz: La instrucción para descubrir todas las guacas del Piru y sus camayos y haziendas. *Journal de la Société des Américanistes* 56: 7–39.

Duvoils, Pierre. 1976. La capacocha. Mecanismo y función del sacrificio humano, su proyección geométrica, su papel en la política integracionista y en la economía redistributiva del Tawantisuyu. *Allpanchis* 9: 11–57.

Earle, Duncan. 2008. Maya Caves across Time and Space. In *Pre-Columbian Landscapes of Creation and Origin*, edited by John E. Staller, pp. 67–93. New York: Springer.

Eliade, Mircea. 1964. *Shamanism: Archaic Techniques of Ecstasy*. Translated by W. R. Trask. Bollingen Series Volume 76. New York: Bollingen Foundation.

Eliade, Mircea, and Lawrence E. Sullivan. 1987. Center of the World. In *The Encyclopedia of Religion*, edited by Mircea Eliade, pp. 166–171. New York: Macmillan Free Press.

Elliot, J. H. 1963. *Imperial Spain: 1468–1716*. New York: Mentor Books.

Espinosa Soriano, Waldemar. 1988a. *Ethnohistoria ecuatoriana: Estudios y documentos*. Quito: Abya-Yala.

Espinosa Soriano, Waldemar. 1988b. *Los cayambes y caranques: Siglos XV–XVI: El testimonio de la ethnohistoria*. Tomo I. Quito: Instituto Otavaleño de Antropología.

Fabian, Johannes. 1983. *Time and the Other: How Anthropology Makes Its Object*. New York: Columbia University Press.

Fernández, Carla. 2006. *Taller Flora*. Mexico City: Editorial Diamantina SA de CV.

Fitzsimmons, Janet. 2005. Pre-Hispanic Rain Ceremonies in Blade Cave, Sierra Mazateca, Oaxaca, Mexico. In *In the Maw of the Earth Monster: Mesoamerican Ritual Cave Use*, edited by J. Brady and K. Prufer, pp. 91–116. Austin: University of Texas Press.

Fitzsimmons Steele, Janet. 1997. Cave Rituals in Oaxaca, Mexico. Paper presented at the Convention of the Society for American Archaeology, Nashville, TN, April 3.

Fletcher, Richard A. 1984. *Saint James's Catapult: The Life and Times of Diego Gelmírez of Santiago de Compostela*. Oxford: Clarendon Press.

Flores Ochoa, Jorge A. 1977. Enqa, enqaychu, illa y khuya rumi. In *Pastores de Puna*, edited by Jorge A. Flores Ochoa, pp. 211–237. Lima: Instituto de Estudios Peruanos.

Florescano, Enrique. 1999. *El Mito de Quetzalcoatl*. México, DF: Fondo de Cultura Económica.

Fonvielle, Wilfrid. 1875. *Thunder and Lightning*. Translation by Thomas Lamb Phipson. New York: Scribner, Armstrong and Company.

Foster, George. 1944. *Nagualism* in Mexico and Guatemala. *Acta Americana* 2: 85–103.

Foster, George. 1945. *Sierra Popoluca Folklore and Beliefs*. University of California Publications on American Archaeology and Ethnology, Vol. 42, No. 2. Berkeley.

Foster, Lynn V. 2005. *Handbook to Life in the Ancient Maya World*. New York: Oxford University Press.

Fox, James A., and John S. Justeson. 1984. Polyvalence in Mayan Hieroglyphic Writing. In *Phoneticism in Mayan Hieroglyphic Writing*, edited by J. Justeson and L. Campbell, pp. 17–76. Publication No. 9. Albany: Institute for Mesoamerican Studies, State University of New York.

Frazer, Sir James George. 1900 [1890]. *The Golden Bough: A Study in Magic and Religion*. 2nd ed. (in 3 vols.), vol. 3. London & New York: Macmillan.

Frazer, Sir James G. 1926. *The Worship of Nature*. London: Macmillan & Co.

Frazer, James George. 2000 [1890]. *The Golden Bough: A Study in Magic and Religion*. Part 1, *The Magic Art and the Evolution of Kings*, vol. 1. London: Adamant Media Corp.

Freidel, David, Linda Schele, and Joy Parker. 1993. *Maya Cosmos*. New York: William Morrow.

Fresco, Antonio 1983. Arquitectura de Ingapirca (Cañar-Ecuador). *Miscelánea Antropológica Ecuatoriana* 3: 195–217. Guayaquil, Ecuador.

Friedman, John S. 2008. *Out of the Blue*. New York: Random House.

Furst, Jill Leslie McKeever. 1998. The Nahualli of Christ: The Trinity and the Nature of the Soul in Ancient Mexico. *RES: Anthropology and Aesthetics* 33: 208–224.

Gade, Daniel. 1983. Lightning in the Folklife and Religion of the Central Andes. *Anthropos* 78: 770–787.

Garcia Quintanilla, Julio. 1963. *Historia de la iglesia de los Charcas o la Plata*. 4 vols. Sucre.

Garcilaso de la Vega, "El Inca." 1961 [1609]. *The Incas: Royal Commentaries of the Incas*. Edited by Alain Gheerbrant, translated by Maria Jolas. New York: Avon Books, Discus Edition.

Garcilaso de la Vega, "El Inca." 1963 [1609]. *The Incas: Royal Commentaries of the Incas*. Edited by Alain Gheerbrant, translated by Maria Jolas. New York: Avon Books, Discus Edition.

Garcilaso de la Vega, "El Inca." 1966 [1609]. *The Incas: Royal Commentaries of the Incas*. Parts 1–2, edited by Alain Gheerbrant, translated by Harold V. Livermore. Austin: University of Texas Press.

Gelles, Paul H. 1995. Equilibrium and Extraction: Dual Organization of the Central Andes. *American Ethnologist* 22 (4): 710–742.

Gelles, Paul H. 1996. Introduction to *Andean Lives*, by Gregorio Condori Mamani and Asunta Quispe Huamán. Originally edited by Ricardo Valderrama and Carmen Escalante, pp. 1–13; translated from the Quechua and with annotations and revised glossary by Paul H. Gelles and Gabriela Martínez Escobar. Austin: University of Texas Press.

Gentry, Alwyn H. 1977. Endemism in Tropical versus Temperate Plant Communities. In *Conservation Biology*, edited by M. E. Soulé, pp. 153–181. Sunderland, MA: Sinauer Associates.

Girard, Rafael. 1995. *People of the Chan*. Chino Valley, AZ: Continuum Foundation.

Glass-Coffin, Bonnie. 1998. *The Gift of Life: Female Spirituality and Healing in Northern Peru*. Albuquerque: University of New Mexico Press.

Gonzalez, Roberto J. 2001. *Zapotec Science: Farming and Food in the Northern Sierra of Oaxaca*. Austin: University of Texas Press.

González Holguin, Diego. 1989. Vocabulario de la lengua general de todo el Peru llamado Qquichua o del Inca [1608, 1952]. Lima: Universidad Nacional Mayor de San Marcos.

Gonzalez Suárez, Fredrico. 1878. *Estudio histórico sobre los Cañaris, antiquos habitantes de la Provincia del Azuay en la republica del Ecuador*. Quito: Casa la Cultura Ecuatoriana.

Gonzalez Suárez, Fredrico. 1892. *Atlas arqueológico ecuatoriano*. Quito: Casa la Cultura Ecuatoriana.

Gose, Peter. 1994. *Deathly Waters and Hungry Mountains: Agrarian Ritual and Class Formation in an Andean Town*. Toronto: University of Toronto Press.

Gose, Peter. 1996. Oracles, Divine Kingship, and Political Representation in the Inka State. *Ethnohistory* 43 (1): 1–32.

Gossen, Gary H. 1972. Temporal and Spatial Equivalents in Chamula Ritual Symbolism. In *Reader in Comparative Religion: An Anthropological Approach*, 3rd ed. New York: Harper and Row.

Govers, Cora. 2006. *Performing the Community: Representation, Ritual and Reciprocity in the Totonac Highlands of Mexico*. Berlin: Lit Verlag.

Graber, J., W. Ummenhofer, and H. Herion. 1996. Lightning Accident with Eight Victims: Case Report and Brief Review of Literature. *Journal of Trauma* 40: 288–290.

Greenberg, James B. 1981. *Santiago's Sword: Chatino Peasant Religion and Economics*. Berkeley: University of California Press.

Greenlee, Robert F. 1944. Medicine and Curing Practices of the Modern Florida Seminoles. *American Anthropologist* 46: 317–328.

Groark, Kevin P. 2010. The Angel in the Gourd: Ritual, Therapeutic, and Protective Uses of Tobacco (Nicotiana Tabacum) among the Tzeltal and Tzotzil Maya of Chiapas, Mexico. *Journal of Ethnobiology* 30: 5–30.

Grove, David. 1970. *The Olmec Paintings of Oxtotitlan Cave, Guerrero, Mexico*. Studies in Pre-Columbian Art and Archaeology, No. 6. Washington, DC: Dumbarton Oaks.

Guaman Poma de Ayala, Felipe (Waman Puma). 1980 [1583–1615]. *Nueva cronica y buen gobierno*, edited by John V. Murra and Rolena Adorno. México, DF: Siglo Veintiuno.

Guinn, J. M. 1905. Rain and Rainmakers. In *Pioneers of Los Angeles County: Southern California Quarterly*, vol. 6, Historical Society of Southern California, Los Angeles, pp. 171–176. Repr., Charleston, SC: Nabu Press, 2010.

Guiteras-Holmes, Calixta. 1961. *Perils of the Soul: The Worldview of a Tzotzil Indian*. New York: Free Press of Glencoe.

Guthrie, Stewart Elliott. 1993. *Faces in the Clouds: A New Theory of Religion*. New York: Oxford University Press.

Heggarty, Paul. 2007. Linguistics for Archaeologists: Principles, Methods and the Case of the Incas. *Cambridge Archaeological Journal* 17 (3): 311–340.

Heggarty, Paul. 2008. Linguistics for Archaeologists: A Case Study in the Andes. *Cambridge Archaeological Journal* 18 (1): 35–56.

Heggarty, Paul, and David Beresford-Jones. 2010. Agriculture and Language Dispersals: Limitations, Refinements and an Andean Exception? *Current Anthropology* 51 (2): 163–191.

Hemming, John. 1970. *The Conquest of the Incas*. San Diego: Harcourt.

Hemming, John, and Edward Ranney. 1982. *Monuments of the Incas*. Boston: Little, Brown.

Hermitte, Esther. 1964. Supernatural Power and Social Control in a Modern Maya Village. PhD diss., Department of Anthropology, University of Chicago.

Hermitte, Esther. 1970. El concepto del *Nahual* entre los Mayas de Pinola. In *Ensayos de Antropología en la Zona Central,* edited by N. A. McQuown and J. Pitt-Rivers. México, DF: Institutio National Indigenista.

Hernández Príncipe, Rodrigo. 1923 [1622]. Mitologia andina. *Inca* 1: 25–68.

Heyden, Doris. 1975. An Interpretation of the Cave underneath the Pyramid of the Sun in Teotihuacan, Mexico. *American Antiquity* 40 (2): 131–147.

Hocquenghem, Anne-Maria. 1993a. Bases del Intercambio entre lãs sociedades norperuanos y surecuatorianas: Una zona de transición entre 1500 A.C. y 600 D.C. *Bulletin del'Institut Français d'Etudes Andines* 22 (2): 443–466.

Hocquenghem, Anne-Maria. 1993b. Rutas de entrada del *mullu* en el extremo norte del Perú. *Bulletin del'Institut Français d'Etudes Andines* 22 (3): 701–719.

Hocquenghem, Anne-Maria. 1994. Los Españoles en los caminos del extremo norte del Peru en 1532. *Bulletin del'Institut Français d'Etudes Andines* 23 (1): 1–67.

Hofling, Charles Andrew. 1997. *Itzaj Maya-Spanish-English Dictionary.* Salt Lake City: University of Utah Press.

Hoogshagen Noordsy, Searle, and Hilda Halloran de Hoogshagen. 1993. *Diccionario Mixe de Coatlán, Oaxaca.* México, DF: Instituto Lingüistico de Verano, AC.

Hosler, Dorothy. 2009. West Mexican Metallurgy: Revisited and Revised. *Journal of World Prehistory* 22: 185–212.

Hosler, Dorothy, Heather Lechtman, and Olaf Holm. 1990. *Axe-Monies and Their Relatives.* Studies in Pre-Columbian Art and Archaeology, 30. Washington, DC: Dumbarton Oaks Research Library and Collection.

Houston, Stephen D., and David Stuart. 1989. *The Way Glyph: Evidence for "Co-Essences" among the Classic Maya.* Research Reports on Ancient Maya Writing, No. 30. Washington, DC: Center for Maya Research.

Houston, Stephen, Oswaldo Chinchilla Mazariegos, and David Stuart. 2001. *The Decipherment of Ancient Maya Writing.* Norman: University of Oklahoma Press.

Huber, Brad. 1990. The Recruitment of Nahua Curers: Role Conflict and Gender Brad R. *Ethnology* 29 (2): 159–176.

Hunn, Eugene S. 2008. *A Zapotec Natural History.* Tucson: University of Arizona Press.

Hurtado de Mendoza, D. Diego 1976 [1560]. *Crónica de la Provincia de San Antonio de los Charcas.* La Paz.

Hyland, Sabine. 2003. *The Jesuit and the Incas: The Extraordinary Life of Padre Blas Valera, S.J.* Ann Arbor: University of Michigan Press.

Hyslop, John. 1990. *Inca Settlement Planning.* Austin: University of Texas Press.

Idrovo-Urigüen, Jaime. 2000a. El Formativo el la Sierra Ecuatoriana. In *Formativo Sudamericano: Un Revaluación,* edited by Paulina Ledergerber-Crespo, pp. 114–123. Quito: Abya Yala.

Idrovo-Urigüen, Jaime. 2000b. *Tomebamba: Arqueología e historia de una cuidad imperial.* Cuenca: Ediciones del Banco Central del Ecuador.

Ingham, John M. 1986. *Mary, Michael and Lucifer: Folk Catholicism in Central Mexico.* Austin: University of Texas Press.

Ions, Veronica. 1967. *Indian Mythology.* London: Paul Hamlyn.

Isbell, Billie Jean. 1978a. *To Defend Ourselves: Ecology and Ritual in an Andean Village*. Austin: University of Texas Press.

Isbell, William H. 1974. Ecología de la expansión de los quechua hablantes. *Revista del Museo Nacional* 34: 143–197.

Isbell, William H. 1978b. Cosmological Order Expressed in Prehistoric Ceremonial Centers. *Proceedings of the Forty-second International Congress of Americanists* 4: 269–297.

Iwaniszewski, Stanislaw. 2001. Reflexiones en torno a los graniceros, planetnicy y renuberos, *Estudios de Cultura Náhuatl* 34: 391–423.

Jijon y Caamaño, Jacinto. 1920. Los tincullpas y notas acerca de la metalurgía de los aborígenes del Ecuador. *Boletín de la Sociedad Ecuatoriana de Estudios Históricos* 5, Nos. 13–14: 4–45.

Jiménez Moreno, Wigiberto. 1970. Mesoamerica before the Toltecs. In *Ancient Oaxaca: Discoveries in Mexican Archeology and History*, edited by John Paddock and Ignacio Bernal, pp. 1–77. Stanford, CA: Stanford University Press.

Joraleman, Donald, and Douglas Sharon. 1993. *Sorcery and Shamanism: Curanderos and Clients in Northern Peru*. Salt Lake City: University of Utah Press.

Josserand, Kathryn, and Nicholas A. Hopkins. 2007. Tila y su Cristo negro: Historia, peregrinación y devoción en Chiapas, Mexico. *Mesoamérica* 49: 82–113.

Joyce, Arthur A. 2009. *Mixtecs, Zapotecs, and Chatinos: Ancient Peoples of Southern Mexico*. New York: Wiley-Blackwell.

Kallen, Stuart A. 1991. *World of the Bizarre*. Edina, MN: Abdo.

Karttunen, Frances. 1983. *An Analytical Dictionary of Nahuatl*. Austin: University of Texas Press.

Kauffman Doig, Federico. 1976. *El Perú Arqueológico*. Lima: Kompaktos.

Kaufman, Terrence. 2003. A Preliminary Mayan Etymological Dictionary. Unpublished ms.

Kaufman, Terrence, and William Norman. 1985. An Outline of Proto-Cholan Phonology, Morphology and Vocabulary. In *Phoneticism in Mayan Hieroglyphic Writing*, edited by J. Justeson and L. Campbell, pp. 77–166. Albany: Institute for Mesoamerican Studies, State University of New York.

Keen, A. Myra. 1971. *Sea Shells of Tropical Western America: Marine Mollusks from Baja California to Peru*. 2nd ed. Stanford, CA: Stanford University Press.

Kehoe, Alice Beck. 2000. *Shamans and Religion: An Anthropological Exploration in Critical Thinking*. Long Grove, IL: Waveland.

Kirchhoff, Paul. 1968. Mesoamerica: Its Geographic Limits, Ethnic Composition and Cultural Characteristics. In *Heritage of Conquest*, edited by Sol Tax, pp. 17–30. New York: Cooper Square.

Klein, Cecelia F., Eulogio Guzmán, Elisa C. Mandell, and Maya Stanfield-Mazzi. 2002. The Role of Shamanism in Mesoamerican Art: A Reassessment. *Current Anthropology* 43 (3): 383–414.

Krappe, Alexander Haggerty. 1932. Spanish Twin Cults. *Studi e Materiale di Storia delle Religioni* 8: 1–22.

Krickeberg, Walter. 1968. Mesoamerica. In *Pre-Columbian American Religions*, edited by W. Krickeberg, H. Trimhorn, W. Müller, and O. Zerries, pp. 5–76. New York: Holt Rinehart, and Winston.

Lane, Belden C. 2002. *Landscapes of the Sacred: Geography and Narratives*. Baltimore: Johns Hopkins University Press.

Laufer, Berthold. 1917. Origin of the Word Shaman. *American Anthropologist* 19: 361–371.

Laughlin, Robert M. 1975. *The Great Tzotzil Dictionary of San Lorenzo Zinacantán*. Washington, DC: Smithsonian Institution Press.

Layard, Austen Henry. 1853. *Monuments of Nineveh*. London: J. Murray.

Lechtman, Heather. 1976. A Metallurgical Site Survey in the Peruvian Andes. *Journal of Field Archaeology* 3: 1–42.

Lechtman, Heather. 1984. Andean Value Systems and the Development of Prehistoric Metallurgy. *Technology and Culture* 25 (1): 1–36.

Lechtman, Heather. 1988. Traditions and Styles in Central Andean Metal-working. In *The Beginning of the Use of Metals and Alloys*, edited by Robert Maddin, pp. 344–378. Cambridge, MA: MIT Press.

Lechtman, Heather. 1999. Afterword to *The Social Dynamics of Technology. Practice, Politics, and World Views*, edited by M. Dobres and C. R. Hoffman, pp. 223–232. Washington, DC: Smithsonian Institution Press.

Lévi-Strauss, Claude. 1966. *The Savage Mind*. Chicago: University of Chicago Press.

Lipp, Frank. 1991. *The Mixe of Oaxaca*. Austin: University of Texas Press.

Lippi, Ronald D., and Alejandra M. Gudino. 2010a. Inkas and Yumbos and Palmitopamba in Northwestern Ecuador. In *Distant Provinces in the Inka Empire: Toward a Deeper Understanding of Inka Imperialism*, edited by M. A. Malpass and Sonia Alconini, pp. 260–278. Iowa City: University of Iowa Press.

Lippi, Ronald D., and Alejandra M. Gudino. 2010b. Palmitopamba: Yumbos e incas en el bosque tropical al noroeste de quito (Ecuador). *Bulletin de l'Institut Français d'Etudes Andines* 39 (3): 623–640.

Lira, Jorge. 1946. *Farmacopea tradicional indigena y practices rituales*. Lima: Talleres Gráficos "el Condor."

Looper, Matthew G. 2003. *Lightning Warrior: Maya Art and Kingship at Quirigua*. Austin: University of Texas Press.

López Austin, Alfredo. 1997. *Tamoanchan, Tlalocan: Places of Mist*. Boulder: University Press of Colorado.

López Austin, Alfredo. 2002. Indigenous Mythology from Present Day Mexico. In *Native Religions and Cultures of Central and South America*, edited by Lawrence E. Sullivan, pp. 33–65. New York: Continuum.

Lorente Fernández, David. 2009. Graniceros, los ritualistas del rayo en México: Historia y etnografía. *Cuicuilco* 47: 201–223.

Lowy, Bernard. 1974. *Amanita muscari*a and the Thunderbolt Legend in Guatemala and Mexico. *Mycologia* 66: 188–191.

Lunardi, F. 1946. El rayo y su culto en los andes. *Revista de Historia* 11: 215–243.

MacCormack, Sabine. 1991. *Religion in the Andes*. Princeton, NJ: Princeton University Press.

Madariaga, Salvador de. 1947. *The Rise of the Spanish American Empire*. New York: Free Press.

Madsen, William. 1955. Shamanism in Mexico. *Southwestern Journal of Anthropology* 11 (1): 48–57.

Magli, Guilo. 2005. Mathematics, Astronomy and Sacred Landscape in the Inka Heartland. *Nexus Network Journal* 7 (2): 22–32.

Malmström, Vincent H. 1995. Geographical Origins of the Tarascans. *Geographical Review* 85: 31–40.

Manca, Maria Cristina. 1995. De las cuevas hasta el cielo pasando a través de los colores de las enfermedades. In *Annuario IEI V*, edited by Fernandez-Galán Rodriguez et al., pp. 223–259. Instituto de Estudios Indígenas, Universidad Autónomo de Chiapas, San Cristóbal de las Casas, Mexico.

Manelis de Klein, Harriet E. 1973. Los Urus, el extraño pueblo de altiplano. *Estudios Andinos* 7: 129–149.

Marcus, Joyce. 1978. Archaeology and Religion: A Comparison of the Zapotec and Maya. *World Archaeology* 10 (2): 172–191.

Marder, William. 2005. *Indians in the Americas: The Untold Story*. New York: Book Tree.

Mariscotti de Görlitz, Ana María. 1973. La posición del señor de los fenómenos meteorológicos en los panteones regionales de los Andes Centrales. *Historia y Cultura* 6: 207–215.

Mariscotti de Görlitz, Ana María. 1978. Los curi y el rayo. *Actes du XLII Congrés international des Américanistes* (Paris 1976), vol. 4: 365–376.

Markman, Roberta H., and Peter T. Markman. 1989. *Masks of the Spirit: Image and Metaphor in Mesoamerica*. Berkeley: University of California Press.

Matienzo Juan de. 1967 [1567]. *Gobierno del Perú*. Traveux de l'Institut Français d'Etudes Andines, Vol. 11.

Maybury-Lewis, David. 1967. *Akwe Shavante Society*. Oxford: Clarendon Press.

Maybury-Lewis, David. 1989. The Quest for Harmony. In *The Attraction of Opposites: Thought and Society in the Dualistic Mode*, edited by David Maybury-Lewis and Uri Almagor, pp. 1–18. Ann Arbor: University of Michigan Press.

Mayr, Ernst. 1979. *Animal Species and Evolution*. Cambridge, MA: Harvard University Press.

McAtee, Waldo L. 1917. Showers of Organic Matter. *Monthly Weather Review* 45 (May): 217. Originally presented to the Biological Society of Washington, January 27, 1917.

McEwan Colin, and Maarten Van de Guchte. 1992. Ancestral Time and Sacred Space in Inca State Ritual. In *The Ancient Americas: Art from Sacred Landscapes*, edited by Richard F. Townsend, pp. 359–371. Chicago: Art Institute of Chicago.

Meddens, Frank M., Nicholas P. Branch, and Cirilo Vivianco Pomacanchari. 2008. High Altitude *Ushnu* Platforms in the Department of Ayacucho Peru, Structure, Ancestors and Animating Essence. In *Pre-Columbian Landscapes of Creation and Origin*, edited by John E. Staller, pp. 315–355. New York: Springer.

Middendorf, E. W. 1895. *Peru: Beobachtungen und Studien über das Land und seine Bewohner*. 3 vols. Berlin.

Miller, Mary Ellen, and Karl A. Taube. 1993. *The Gods and Symbols of Ancient Mexico and the Maya: An Illustrated Dictionary of Mesoamerican Religion*. New York: Thames and Hudson.

Miller, Walter. 1956. *Cuentos Mixes*. Mexico City: Instituto Nacional Indigenista.

Minnaert, M. 1954. *The Nature of Light and the Color of Open Air*. New York: Dover.

Mishkin, Bernard. 1940. Cosmological Ideas among the Indians of the Southern Andes. *Journal of American Folklore* 53: 225–241.

Mishkin, Bernard. 1946. The Contemporary Quechuas. In *Handbook of South American Indians*, vol. 2, edited by Julian H. Steward, pp. 411–470. Washington, DC: Bureau of American Ethnology Smithsonian Institution.

Molina, Cristóbal de. 1959 [1575]. *Ritos y fábulas de los Incas.* Buenas Aires.

Molina, Cristóbal de (El Almagista). 1968 [1552]. Relación de muchas cosas acaescides en el Perú. In *Crónicas Peruanas de Interés Indígena*, edited by Francisco Steve Barba, pp. 56–95. Biblioteca de Autores Españoles, Vol. 209. Madrid: Real Academia Española.

Monaghan, John. 1995. *The Covenants with Earth and Rain: Exchange, Sacrifice, and Revelation in Mixtec Sociality.* Norman: University of Oklahoma Press.

Montejo, Victor. 1984. *El Kanil: Man of Lightning.* Translated by Wallace Kaufman. Carrboro, NC: Signal Books.

Montgomery, John. 2002. *Dictionary of Maya Hieroglyphs.* New York: Hippocrene Books.

Moore, Jerry D. 2005. *Cultural Landscapes in the Ancient Andes: Archaeologies of Place.* Gainesville: University of Florida Press.

Moore, Robert L. 1984. Space and Transformation in Human Experience. In *Anthropology and the Study of Religion*, edited by Robert L. Moore and Frank E. Reynolds, pp. 126–143. Chicago: Center for the Scientific Study of Religion.

Morris, Craig. 1979. Maize Beer in the Economics, Politics, and Religion of the Inca State. In *The Role of Fermented Beverages in Nutrition*, edited by C. Gastineau, W. Darby, and T. Turner, pp. 21–34. New York: Academic Press.

Morris, Craig. 1993. The Wealth of a Native American State: Value, Investment, and Mobilization in Inka Economy. In *Configurations of Power*, edited by John S. Henderson and Patricia J. Netherly, pp. 36–50. Ithaca, NY: Cornell University Press.

Morris, Craig. 2004. Enclosures of Power: The Multiple Spaces of Inca Administrative Palaces. In *Palaces of the Ancient New World*, edited by Susan T. Evans and Joanne Pillsbury, pp. 299–324. Washington, DC: Dumbarton Oaks Museum Research Library and Collection.

Morris, Craig, and Donald E. Thompson. 1985. *Huánuco Pampa: An Inca City and Its Hinterlands.* New York: Thames and Hudson.

Morris, Walter F., Jr. 1987. *Living Maya.* New York: Henry N. Adams.

Murra, John V. 1946. The Tribes of Ecuador. In *Handbook of South American Indians*, vol. 2, edited by Julian H. Steward, pp. 785–822. Bulletin 143, Bureau of American Ethnology. Washington, DC: Smithsonian Institution.

Murra, John V. 1975a [1972]. El control vertical de un máximo de pisos ecológicos en los sociedades andinas. In *Formaciones Económicas y Politicas del Mundo Andino*, edited by J. V. Murra, pp. 59–115. Lima: Instituto de Estudios Peruanos.

Murra, John V. 1975b. El trafico de mullu en la costa del Pacifico. In *Formaciones económicas y politicas del mundo andino*, pp. 255–267. Lima: Instituto de los Estudios Peruanos.

Murra, John V. 1985a. El Archipélago Vertical Revisited. In *Andean Ecology and Civilization: An Interdisciplinary Perspective on Andean Ecological Complementarity*, edited by Shozo Masuda, Izumi Shimada, and Craig Morris, pp. 3–13. Papers from the Wenner-Gren Foundation for Anthropological Research Symposium No. 91. Tokyo: University of Tokyo Press.

Murra, John V. 1985b. The Limits and Limitations of the "Vertical Archipelago" in the Andes. In *Andean Ecology and Civilization: An Interdisciplinary Perspective on Andean Ecological Complementarity*, edited by Shozo Masuda, Izumi Shimada, and Craig Morris, pp. 15–20. Papers from the Wenner-Gren Foundation for Anthropological Research Symposium No. 91. Tokyo: University of Tokyo Press.

Murúa, Martín de. 1922 [1590]. *Historia de los Incas: Reyes del Peru. Colección de libros y documentos referents a la historia del Perú*, 2nd series, 4:1–253, edited by H. Urteaga and C. Romero. Lima.

Nuñez del Prado, Juan Víctor. 1974. The Supernatural World of the Quechua of Southern Peru as Seen from the Community of Qotobamba. In *Native South Americans: Ethnology of the Least Known Continent*, edited by Patricia J. Lyon, pp. 238–251. Prospect Heights, IL: Waveland.

Nuñez del Prado, Oscar. 1973. *Kuyo Chico*. Chicago: University of Chicago Press.

Nuttall, Zelia. 1891. *The Atlatl or Spear-Thrower of the Ancient Mexicans*. Archaeological and Ethnological Papers of the Peabody Museum, Harvard University, Vol. 1, No. 3. Cambridge, MA: Peabody Museum.

Nuttall, Zelia. 1897. Ancient Mexican Superstitions. *Journal of American Folklore* 10: 265–281.

Ortiz, Elio, Antonio Mendez, Alejo Zarzycki, and Janis Alcorn. 2008. Fox Walker on the Parapeti River, Bolivia: The Origins of How the Guarani Live in Íví. In *Pre-Columbian Landscapes of Creation and Origin*, edited by John E. Staller, pp. 161–202. New York: Springer.

Orville, R. E. 1990. Peak Current Variations of Lightning Return Strokes as a Function of Latitude. *Nature* 343: 159–161.

Pachacuti Yamqui, Jorge de Santacruz. 1950 [c. 1613]. Relación de antigüedades desde reyno del Piru. In *Tres relaciones de antiqüedades peruanas*, edited by M. Jiménes de la Espalda, pp. 205–281. Asunción, Paraguay: Editorial Guaraní.

Parsons, Elsie Clews. 1936. *Mitla, Town of Souls*. Chicago: University of Chicago Press.

Paul, Benjamin D., and Clancy McMahon. 2001. Mesoamerican Bonesetters. In *Mesoamerican Healers*, edited by Brad R. Huber and Alan R. Sandstrom, pp. 243–269. Austin: University of Texas Press.

Paulsen, Alice C. 1974. The Thorny Oyster and the Voice of God: *Spondylus* and *Strombus* in Andean Prehistory. *American Antiquity* 39: 597–607.

Paz, Josef. 1995. The Vicissitude of the Alter Ego Animal in Mesoamerica: An Ethnohistorical Reconstruction of Tonalism. *Anthropos* 90 (4/6): 445–465.

Pease, Franklin. 1973. *El Dios Creador Andino*. Lima: Mosca Azul Editoriales.

Phillips, Charles, and David M. Jones. 2005. *Lost History of Aztec and Maya*. New York: Ultimate Editions.

Pillsbury, Joanne. 1996. The Thorny Oyster and the Origins of Empire: Implications of Recently Uncovered *Spondylus* Imagery from Chan Chan, Peru. *Latin American Antiquity* 7 (4): 313–340.

Pillsbury, Joanne. 2004. The Concept of the Palace in the Andes. In *Palaces of the Ancient New World*, edited by Susan T. Evans and Joanne Pillsbury, pp. 181–190. Washington, DC: Dumbarton Oaks Museum Research Library and Collection.

Pillsbury, Joanne, and Susan Toby Evans. 2004. Palaces of the Ancient New World: An Introduction. In *Palaces of the Ancient New World*, edited by Susan T. Evans and Joanne Pillsbury, pp. 1–6. Washington, DC: Dumbarton Oaks Museum Research Library and Collection.

Pinto, Osmar, Jr. 2009a. *Lightning in the Tropics: From a Source of Fire to a Monitoring System of Climatic Changes*. New York: Nova.

Pinto, Osmar, Jr. 2009b. Lightning in the Tropics: A Review. Presented at the Tenth International Symposium on Lightning Protection. November 9–13, 2009, Curitiba, Brazil.

Pinto, O., Jr., K. P. Naccarato, I. R. C. A. Pinto, W. A. Fernandes, and O. P. Neto. 2006. Monthly Distribution of Cloud-to-Ground Lightning Flashes as Observed by Lightning Location Systems. *Geophysical Research Letters* 33, L09811doi:10.1029/2006GL026081.

Piper, Ross. 2007. *Extraordinary Animals: An Encyclopedia of Curious and Unusual Animals.* New York: Greenwood.

Pitarch Ramón, Pedro. 1998. En mitad del cielo: Una oración chamánica tzeltal dirigida a defender un nahual-rayo. *Revista Española de Antropología Americana* 28 (2): 15–241.

Pitt Rivers, Julian. 1970. Spiritual Power in Central America: The Naguals of Chiapas. In *Witchcraft, Confessions, and Accusations*, edited by Mary Douglas, pp. 183–206. London: Tavistock.

Pizarro, Pedro. 1978 [1571]. *Relación del discubrimiento y conquista del Perú*. Lima: Pontífica Universidad Católica del Perú.

Platt, Tristan. 1997. The Sound of Light: Emergent Communication through Quechua Shamanic Dialogue. In *Creating Context in Andean Cultures*, edited by Rosaleen Howard-Malverde, pp. 196–226. Oxford Studies in Anthropological Linguistics. New York: Oxford University Press.

Platt, Tristan. 2002. Without Deceit or Lies: Variable Chinu Readings during a Sixteenth-Century Tribute-Restitution Trial. In *Narrative Threads: Accounting and Recounting in Andean Khipu*, edited by Jeffrey Quilter and Gary Urton, pp. 225–265. Austin: University of Texas Press.

Platt, Tristan, and Pablo Quisbert. 2007. Knowing Silence and Merging Horizons: The Case of the Great Potosí Cover Up. *Ways of Knowing: New Approaches in the Anthropology of Experience and Learning*, edited by Mark Harris, pp. 113–138. Oxford: Berghahn.

Polo de Ondegardo, Juan. 1916 [1561]. *Los errors y supersticiones populares de Bolivia.* 3rd ed. La Paz.

Pool, Christopher A. 2007. *Olmec Archaeology and Early Mesoamerica*. Cambridge: Cambridge University Press.

Rakov, Vladimir A., and Martin A. Uman. 2003. *Lightning: Physics and Effects*. Cambridge: Cambridge University Press.

Ramirez del Aguila, Pedro. 1978. *Noticias políiticas de Indias*, edited by Jaime Urioste. Sucre: Arano.

Ramos Gómez, Oscar Gerardo. 1988. *Sebastián de Benalcázar conquistador de Quito y Popayan*. Madrid: Biblioteca Iberoamericana; Ediciones Anaya SA.

Ravicz, Robert S. 1965. *Organization social de los Mixtecos*. México, DF: Instituto Nacional Indigenista.

Read, Kay Almere, and Jason K. González. 2002. *Mesoamerican Mythology: A Guide to the Gods, Heroes, Rituals, and Beliefs of Mexico and Central America*. New York: Oxford University Press.

Recinos, Adrian, Goetz Delia, and Sylvanus G. 1950. *Popol Vuh: The Sacred Book of the Ancient Quiché Maya*. Norman: University of Oklahoma Press.

Redfield, Robert, and Alfonso Villa Rojas. 1934. *Chan Kom: A Maya Village*. Chicago: University of Chicago Press.

Rehren, Thilo, and Matilda Temme. 1994. Pre-Columbian Gold Processing at Putushio, South Ecuador: The Archaeological Evidence. *Archaeometry of Pre-Columbian Sites and Artifacts*, edited by D. A. Scott and P. Meyers, pp. 267–284. Malibu, CA: Getty Conservation Institute.

Reichel-Dolmatoff, Gerardo. 1972. The Feline Motif in Prehistoric San Agustín Sculpture. In *The Cult of the Feline*, edited by Elizabeth P. Benson, pp. 51–68. Washington, DC: Dumbarton Oaks Research Library and Collection.

Reichel-Dolmatoff, Gerardo. 1975. *The Shaman and the Jaguar: A Study of Narcotic Drugs among the Indians of Colombia*. Philadelphia: Temple University Press.

Reilly, F. Kent, III. 1989. The Shaman in Transformation Pose: A Study of the Theme of Rulership in Olmec Art. *Record of the Art Museum, Princeton University* 48 (2): 4–21.

Reilly, F. Kent, III. 1994. Visions of Another World: Art, Shamanism, and Political Power in Middle Formative Mesoamerica. PhD diss., Institute of Latin American Studies, University of Texas, Austin.

Reilly, F. Kent, III. 2004. *The Lazy-S: A Formative Period Iconographic Loan to Maya Hieroglyphic Writing*, pp. 1–11. http://www.mesoweb.com/pari/publications/RT10/32_LazyS.html.

Reinhard, Johannes. 1983. High Altitude Archaeology and Andean Mountain Gods. *American Alpine Journal* 25: 54–67.

Reinhard, Johannes. 1985. Sacred Mountains: An Ethno-Archaeological Study of High Andean Ruins. *Mountain Research and Development* 5: 299–317.

Reinhard, Johan, and María Constanza Ceruti. 2010. *Inca Rituals and Sacred Mountains: A Study of the World's Highest Archaeological Sites*. Los Angeles: UCLA Cotsen Institute of Archaeology Press.

Reko, Blas Pablo. 1945. *Mitobotánica Zapoteca*. México, DF: General León 9.

Rice, Prudence. 1993. The Status of Latin American Horizons. In *Latin American Horizons*, edited by Prudence Rice, pp. 357–364. Washington, DC: Dumbarton Oaks Library.

Ridgely, Robert S., and Paul J. Greenfield. 2001. *The Birds of Ecuador*, vol. 1: *Status, Distribution, and Taxonomy*. Foreword by F. B. Gill. Ithaca, NY: Cornell University Press.

Roca W., Demetrio. 1966. El sapo, la culebra y la rana en el folklore actual de la pampa de Anta. *Folklore, Revista de Cultura Tradicional* 1 (1): 41–66.

Roe, Peter G. 1982. *The Cosmic Zygote, Cosmology in the Amazon Basin*. New Brunswick, NJ: Rutgers University Press.

Romero Vivas, Isabel Laura. 1999. El simbolismo del agua entre los Totonacos. *La Palabra y el Hombre* 111: 57–78.

Romney, A. Kimball, and Romaine Romney. 1966. *The Mixtecans of Juxtlahuaca, Mexico*. New York: John Wiley & Sons.

Rostworowski de Diez Canseco, María. 1975. Pescadores, artesanos y mercaderes costeños en el Peru prehispánico. *Revista del Museo Nacional* (Lima) 41: 311–349.

Rostworowski de Diez Canseco, María. 1983. *Estructuras andinas del poder: Ideología religiosa y política*. 2nd ed. Lima: Instituto de Estudios Peruanos.

Rostworowski, María, and Craig Morris. 1999. Inka Power and Its Social Foundations. In *The Cambridge History of the Native Peoples of the Americas*, vol. 3: *South America, Part 1*, edited by Frank Salomon and Stuart B. Schwartz, pp. 769–863. Cambridge: Cambridge University Press.

Rowe, John. 1946. Inca Culture at the Time of the Spanish Conquest. In *Handbook of South American Indians*, vol. 2., edited by Julian H. Steward, pp. 43–50. Bulletin 143, Bureau of American Ethnology. Washington, DC: Smithsonian Institution.

Rowe, John. 1974. Linguistic Classification Problems in South America. In *Native South Americans: Ethnology of the Least Known Continent*, edited by Patricia J. Lyon, pp. 238–251. Prospect Heights, IL: Waveland.

Rowe, John. 1978. La fecha de la muerte de Wayna Qhapaq. *Historica* 2 (1): 83–88.

Roys, Ralph L. 1967. *The Book of Chilam Balam of Chumayel*. Norman: University of Oklahoma Press.

Ruiz de Alarcón, Hernando. 1982. *Aztec Sorcerers in Seventeenth Century Mexico: The Teratise on Superstitions by Hernando Ruiz de Alarcon*. Edited and translated by Michael D. Coe and Gordon Whittaker. Monograph Series 7, Institute for Mesoamerican Studies. Albany: State University of New York Press.

Ruz, Mario. 1986. *Domingo de ara vocabulario de lengua tzeldal segújn el orden de copanabastla*. México, DF: Universidad Nacional Autonoma de Mexico.

Ryall, Julian. 2010. Lightning Makes Mushrooms Multiply. *National Geographic Daily News*, April 9. news.nationalgeographic.com/news/2010/04/100409-lightning-mushrooms-japan-harvest/.

Salazar, Ernesto. 1998. Naturalzea y distribución de los montículos precolombinos de Cuenca del Alto Upano, Ecuador. In *Intercambio y comercio entre costa, Andes y selva arqueologia y etnohistoria de suramerica*, edited by F. Cárdenas-Arroyo and T. L. Bray, pp. 185–211. Bogotá: Departamento de Antropología, Universidad de los Andes.

Salazar, Ernesto. 2004. Cuenca y su région: En busca del tiempo perdido. In *Cuenca: Santa Ana de las Aguas*, edited by E. Salazar, J. Martínez, A. Abad, and F. Aguilar, pp. 18–85. Quito: Libri Mundi.

Salazar, Ernesto, Diego Jaramillo, Juan Martínez, Ana Abad, and Felipe Aguilar. 2004. *Cuenca: Santa Ana de las Aguas*. Quito: Libri Mundi.

Saler, Benson. 1964. Nagual, Witch, and Sorcerer in a Quiche Village. *Ethnology* 3: 305–328.

Sallnow, Michael. 1987. *Pilgrims of the Andes: Regional Cults in Cusco*. Washington, DC: Smithsonian Institution Press.

Salomon, Frank L. 1982. Chroniclers of the Impossible: Notes on Three Peruvian Indigenous Historians. In *From Oral to Written Expression*, pp. 9–40. Syracuse, NY: Syracuse University Press.

Salomon, Frank L. 1986. *Native Lords of Quito in the Age of the Incas: The Political Economy of the Northern Andean Chiefdoms*. Cambridge: Cambridge University Press.

Salomon, Frank L. 1987. Ancestors, Grave Robbers, and the Possible Antecedents of Cañari "Inca-ism." In *Natives and Neighbors in South America.*, edited by Herald O. Skar and Frank L. Salomon, pp. 207–232. Ethnologiska Studier 38. Göteborg: Göteborgs Ethnografiska Museum.

Salomon, Frank L., and George L. Urioste. 1991 [c. 1598–1608]. *The Huarochiri Manuscript: A Testament of Ancient and Colonial Andean Religion*. Translated from the Quechua by Frank L. Salomon and George L. Urioste. Austin: University of Texas Press.

Sandstrom, Alan R. 1991. *Corn Is Our Blood*. Norman: University of Oklahoma Press.

San Pedro, Fray Juan de. 1992 [1560]. *Le persecución del demonio: Crónica de los primeros agostinos el norte del Perú*. Melaga and México, DF: Centro Andinoy Mesoamericano de Estudios Interdisciplinarios.

San Pedro, Fray Juan de. 1560. *Relaciones de primeros Agustinos*. Madrid.

Sarmiento de Gamboa, Pedro. 1907 [1572]. *The History of the Incas*, translated by Sir Clements Markham. Cambridge, UK: Haklyut Society.

Saunders, Nicholas J. 1989. *People of the Jaguar: The Living Spirit of Ancient America*. London: Souvenir.

Saville, Marshall H. 1924. *The Gold Treasure of Sigsig, Ecuador*. Leaflets of the Museum of the American Indian No. 3. New York: Heye Foundation.

Schele, Elaine. 2007. Potential Artistic Representations of the Snake Hemipenis in Mesoamerican Art. *FAMSI Journal of the Ancient Americas*. research.famsi.org/aztlan /uploads/papers/schele-snake-hemipenis.pdf, pp. 1–13.

Schellen, Heinrich, and Sir William Huggins. 1872. *Spectrum Analysis in Its Application to Terrestrial Substances and the Physical Constitution of the Heavenly Bodies*. New York: D. Appleton and Company.

Schoenhals, Alvin, and Louise Schoenhals. 1965. *Vocabulario Mixe de Totontepec*. México, DF: Summer Institute of Linguistics.

Schuster, Angela M. H. 1999. Redating the Madrid Codex. *Archaeology* 52 (1): 26–27.

Sharon, Douglas G. 1972. Eduardo the Healer. *Natural History* 81 (9): 32–47.

Sharon, Douglas G. 1976. Becoming a Curandero in Peru. In *Enculturation in Latin America: An Anthology*, edited by Johannes Wilbert, pp. 213–236. Los Angeles: UCLA Latin American Center Publication.

Sharon, Douglas G. 1978. *The Wizard of the Four Winds: A Shaman's Story*. New York: Free Press.

Sharon, Douglas G. 2000. *Shamanism and the San Pedro Cactus*. San Diego Museum Papers 37.

Sharon, Douglas G. 2001. Speculations on Moche Mountain Scenes. In *Mortuary Practices and Ritual Associations: Shamanic Elements in Prehistoric Funerary Contexts in South America*, edited by John E. Staller and Elizabeth J. Currie, pp. 109–116. Oxford: British Archaeological Reports International Series 982.

Sharon, Douglas G., and Christopher B. Donnan. 1977. The Magic Cactus: Ethnoarchaeological Continuity in Peru. *Archaeology* 30: 374–381.

Shaw, Mary. 1971. *According to Our Ancestors*. Norman: Summer Institute of Linguistics of the University of Oklahoma.

Silverman-Proust, Gail. 1988. Weaving Technique and the Registration of Knowledge in the Cuzco Area of Peru. *Journal of Latin American Lore* 14 (2): 207–241.

Slocum, Marianna. 1965. The Origin of Corn, and Other Tzeltal Myths. *Tlalocan* 5: 1–45.

Solá, Donald F. 1967. *Gramatica del Quechua de Huánuco, plan de fomento lingüístico*. Lima: Universidad Nacional Mayor de San Marcos.

Spero, Joanne. 1987. Lightning Men and Water Serpents: A Comparison of Mayan and Mixe-Zoquean Beliefs. Master's thesis, University of Texas at Austin.

Spero, Joanne. 1991. Beyond Rainstorms: The Kawak as an Ancestor, Warrior, and Patron of Witchcraft. Originally published in *Sixth Palenque Round Table, 1986*, edited by Virginia M. Fields, pp. 184–193. Electronic version. Norman: University of Oklahoma Press.

Staller, John E. 2000–2002. Explorando el contexto y el significado del intercambio a larga distancia en un paisaje sagrado: El caso del Formativo Ecuatoriano. *Cuadernos de Historia y Arqueología*, Nos. 54–56: 56–114. Publicación de la Casa la Cultura Ecuatoriana. Benjamin Carrion, Nucleo del Guayas. Guayaquil, Ecuador.

Staller, John E. 2001. Shamanic cosmology embodied in Valdivia VII–VIII Mortuary context from the site of La Emerenciana, Ecuador. In *Mortuary Practices and Ritual Associations: Shamanic Elements in Prehistoric Funerary Contexts in South America*, edited by John E. Staller and Elizabeth J. Currie, pp. 19–36. British Archaeological Reports International Series 982. Oxford: Archaeopress.

Staller, John E. 2003. Donald Collier: A Curator's Life. In *Curators, Collections, and Contexts: Anthropology at the Field Museum, 1893—2002*, edited by Stephen E. Nash and Gary M. Feinman, pp. 189–198. Fieldiana, Anthropology, n.s., No. 36. Chicago: Field Museum of Natural History.

Staller, John E. 2006. The Social, Symbolic and Economic Significance of *Zea mays* L. in the Late Horizon Period. In *Histories of Maize: Multidisciplinary Approaches to the Prehistory, Linguistics, Biogeography, Domestication, and Evolution of Maize*, edited by John E. Staller, Robert H. Tykot, and Bruce F. Benz, pp. 449–467. Walnut Creek, CA: Left Coast.

Staller, John E. 2007. Un reevaluación del papel de la ideología en el intercambio de larga distancia temprano y a los orígenes de la civilización andina. In *II Congreso Ecuatoriano de Antropologia y Arqueologia: Balance de la última década: aportes, retos y nuevos temas*, edited by Fernando García, pp. 511–548. Tomo 1. Quito: Ediciones Abya Yala.

Staller, John E. 2008a. An Introduction to Pre-Columbian Landscapes of Creation and Origin. In *Pre-Columbian Landscapes of Creation and Origin*, edited by John Edward Staller, pp. 1–9. New York: Springer.

Staller, John E. 2008b. Dimensions of Place: The Significance of Centers to the Development of Andean Civilization: An Examination of the *Ushnu* Concept. In *Pre-Columbian Landscapes of Creation and Origin*, edited by John Edward Staller, pp. 269–313. New York: Springer.

Staller, John E. 2010a. *Maize Cobs and Cultures: History of Zea mays L.* Berlin: Springer.

Staller, John E. 2010b. Ethnohistoric Sources on Foodways, Feasts and Festivals in Mesoamerica. In *Pre-Columbian Foodways: Interdisciplinary Approaches to Food, Culture and Markets in Ancient Mesoamerica*, edited by John Edward Staller and Michael Carrasco, pp. 23–69. New York: Springer.

Staller, John E. 2011. San Catequilla de Pichincha: Lightning *Huacas*, Ancient Astronomy, and Inca Expansion. Presented at an invited symposium titled Recent Advances Regarding Ancient Advances of the Incas in Chinchaysuyu at the Seventy-sixth Annual Meeting of the Society for American Archaeology, March 30 to April 3, Sacramento, CA.

Stark, Louisa R. 1985. Indigenous Languages of Lowland Ecuador: History and Current Status. In *South American Indian Languages: Retrospect and Prospect*, edited by Harriet E. Manelis Klein and Louisa R. Stark, pp. 157–193. Austin: University of Texas Press.

Stokstad, Marilyn. 1978. *Santiago de Compostela in the Age of the Great Pilgrimages*. Norman: University of Oklahoma Press.

Stross, Brian. 1973. Social Structure and Role Allocation in Tzeltal Oral Literature. *Journal of American Folklore* 86: 95–113.

Stross, Brian. 2006. Maize in Word and Image in Southeastern Mesoamerica. In *Histories of Maize: Multidisciplinary Approaches to the Prehistory, Linguistics, Biogeography, Domestication, and Evolution of Maize*, edited by John E. Staller, Robert H. Tykot, and Bruce F. Benz, pp. 578–599. Walnut Creek, CA: Left Coast.

Sullivan, Lawrence E. 1984. Sacred Music and Sacred Time. *World of Music* 26 (3): 33–52.

Sullivan, Lawrence E. 1985. Above, Below, or Far Away: Andean Cosmogony and Ethical Order. *Cosmogony and Ethical Order: New Studies in Comparative Ethics*, edited by Robin W. Lovin and Frank Reynolds, pp. 98–129. Chicago: University of Chicago Press.

Sullivan, Lawrence E. 1986. Sounds and Senses: Towards an Understanding of Hermeneutics of Performance. *History of Religions* 29 (1): 1–33.

Sullivan, Lawrence E. 1987. Nature: The Worship of Nature. In *The Encyclopedia of Religion*, edited by Mircea Eliade, pp. 324–328. New York: Macmillan.

Sullivan, Lawrence E. 1988. *Icanchu's Drum: An Orientation to Meaning of South American Religions*. New York: Macmillan.

Sullivan, Lawrence E., ed. 2002. *Native Religions of Central and South America*. London: Continuum.

Sullivan, William. 1996. *The Secret of the Incas: Myth, Astronomy, and the War against Time*. New York: Three Rivers.

Tate, Carolyn. 1992. *Yaxchilan: The Design of a Maya Ceremonial City*. Austin: University of Texas Press.

Taube, Karl A. 1992a. *The Major Gods of Ancient Yucatan*. Washington, DC: Dumbarton Oaks.

Taube, Karl A. 1992b. The Temple of Quetzalcoatl and the Cult of Sacred War at Teotihuacan. *RES: Anthropology and Aesthetics* 21: 53–87.

Taube, Karl A. 1996a. The Olmec Maize God: The Face of Corn in Formative Mesoamerica. *Res—Anthropology and Aesthetics* 29–30: 39–81.

Taube, Karl. 1996b. Lightning Celts and Corn Fetishes: The Formative Olmec and the Development of Maize Symbolism in Mesoamerica and the American Southwest. In *Olmec Art and Archaeology in Mesoamerica*, edited by John E. Clark and Mary E. Pye, pp. 297–337. Washington, DC: National Gallery of Art; New Haven, CT: Yale University Press.

Taube, Karl A. 2000. The Turquoise Hearth: Fire, Self-Sacrifice, and the Central Mexican Cult of War. In *Mesoamerica's Classic Heritage: From Teotihuacan to the Aztecs*, edited by David Carrasco, Lindsay Jones, and Scott Sessions. Boulder: University Press of Colorado.

Taussig, Helen B. 1968. Death from Lightning and the Possibility of Living Again. *Annals of Internal Medicine* 68: 1345–1353.

Taussig, Michael T. 1987. *Shamanism, Colonialism and the Wild Man: A Study in Terror and Healing*. Chicago: University of Chicago Press.

Taylor, Gerald, ed. 1987. *Ritos y Tradiciones de Huarochirí del Siglo XVII*. Lima: Instituto de Estudios Peruanos.

Tedlock, Barbara. 1992. *Time and the Highland Maya*. Rev ed. Albuquerque: University of New Mexico Press.

Tedlock, Dennis. 1996. *Popol Vuh*. New York: Simon and Schuster.

Temme W., Mathilda. 2000. El Formativo en la Sierra Sur del Ecuator. In *Formativo Sudamericano: Un Revaluación*, edited by Paulina Ledergerber-Crespo, pp. 124–138. Quito: Abya Yala.

Thompson, J. Eric S. 1930. *Ethnology of the Mayas of Southern and Central British Honduras*. Field Museum of Natural History Anthropology Series, vol. 17, no. 2. Chicago: Field Museum of Natural History.

Thompson, J. Eric S. 1934. *Sky Bearers: Colors and Direction in Maya and Mexican Religion.* Contributions to American Archaeology 10. Washington, DC: Carnegie Institute of Washington.

Thompson, J. Eric S. 1970. *Maya History and Religion.* Norman: University of Oklahoma Press.

Tie, X. X., R. Zhang, G. Brasseur, L. Emmons, and W. Lei. 2001. Effects of Lightning on Reactive Nitrogen and Nitrogen Reservoir Species in the Troposphere. *Journal of Geophysical Research* 106: 3167–3178.

Toor, Frances. 1947. *A Treasury of Mexican Folkways.* México, DF: Mexico Press.

Topic, John R. 1992. Las Huacas de Huamachuco: Precisions en torno a una imagen indígena de un paisaje andino. In *Fray Juan de San Pedro La Persecución del Demonio: Crónica de los primeros agustinos en el Norte del Perú,* transcript by E. E. Deeds, pp. 41–99. Málaga: Editrorial Algazara.

Topic, John R. 1998. Ethnogenesis in Huamachuco. *Andean Past* 5: 109–127.

Topic, John R., Theresa L. Topic, and Alfredo Melly Cava. 2002. Catequil, the Archaeology, Ethnohistory, and Ethnography of a Major Provincial Huaca. In *Andean Archeology I: Variations in Sociopolitical Organization,* edited by William H. Isbell and Helena Silverman, pp. 303–335. New York: Kluwer Academic and Plenum.

Torero, Alfredo. 1964. Los dialectos quechuas. *Anales Científicos de la Universidad Agraria* 2: 446–478.

Torero, Alfredo. 1972. Linguística e historia de los Andes del Perú y Bolivia. In *El reto del multilingüísmo en el Perú,* edited by Alberto Estobar, pp. 51–106. Serie Perú, Problema no. 9. Lima, Peru: Instituto de Estudios Peruanos.

Torero, Alfredo. 1974. *El Quechua y la historia social andina.* Lima: Fondo Editorial Horizante/Institut Françaís d'Études Andines. (Reprinted in 2007.)

Torero, Alfredo. 1984. El comercio lejano y la difusión del quéchua. El caso del Ecuador. *Revista Andina* 2 (2): 367–402.

Torero, Alfredo. 2002. *Idiomas de los Andes—Linguistica e Historia.* Lima: Editorial Horizonte/Instituto Francais des Estudes Andinas.

Townsend, Richard F., ed. 1992. *The Ancient Americas: Art from Sacred Landscapes.* Chicago: Art Institute of Chicago.

Tschopik, Harry. 1951. *Magia cen Chicuito: Los aymara del peru.* Instituto Indigenista Interamericano Ediciones Especiales 50.

Turner, Paul R. 1972. *The Highland Chontal.* New York: Holt, Rinehart and Winston.

Turner, Paul R., and Shirley Turner. 1971. *Dictionary Chontal to Spanish—English Spanish to Chontal.* Tucson: University of Arizona Press.

Uhle, Max. 1922. Sepulturas ricas de oro en la provincía del Azuay. *Boletín de la Academia Nacional de Historia* 4 (9): 108–114.

Uhle, Max. 1923. *Las ruinas de Tomebamba.* Cuenca: Centro de Estudíos Históricos y Geográficos del Azuay.

Urcid, Javier. 2005. Zapotec Writing: Knowledge, Power, and Memory in Ancient Oaxaca. http://www.famsi.org/zapotecwriting/index.html.

Urton, Gary. 1981. *At the Crossroads of the Earth and the Sky.* Austin: University of Texas Press.

Urton, Gary. 1985. Animal Metaphors and the Life Cycle in an Andean Community. In *Animal Myths and Metaphors in South America,* edited by Gary Urton, pp. 251–284. Salt Lake City: University of Utah Press.

Urton, Gary. 1990. *The History of a Myth: Pacariqtambo and the Origin of the Inkas*. Austin: University of Texas Press.

Urton, Gary. 1992. Communualism and Differentiation in an Andean Community. In *Andean Cosmologies through Time: Persistence and Emergence*, edited by R. V. H. Dover, K. E. Seibold, and J. H. McDowell, pp. 229–260. Bloomington: University of Indiana Press.

Urton, Gary. 1999. *Inca Myths*. Austin: University of Texas Press; London: British Museum.

Valcarcel, Luis E. 1964. *Historía del antiquo Perú*. Lima: Editorial Juan Mejia Baça.

Valderrama, Ricardo, and Carmen Escalante. 1977. *Gregorio Condori Mamani: Autobiografía*. CERA Bartolomé de las Casas. Cuzco: Edición bilingüe quecchua-español.

Valderrama, Ricardo, and Carmen Escalante. 1988. *Del Tata Mallku a la Mama Pacha, riego sociedad y ritos en los Andes Peruanos Desco*. Lima: Centro de Estudios y Promoción del Desarollo.

Verneau, R., and Paul Rivet. 1912. *Etnographic ancienne de l'Equateur*. 2 vols. Mission du Service Geographique de l'Armée pour la Mésure d'un Arc de Méridian Equatorial en Amérique du Sud. Paris: Ministerio de Instrucción Pública.

Villa Rojas, Alfonso. 1947. Kinship and Nagualism in a Tzeltal Community, Southeastern Mexico. *American Anthropologist* 49 (4): 578–587.

Villalba, Fabián. 2011. Arqueología del Valle del Río Quimi. *Arqueología e Historia* 4: 4–13.

Vogt, Evon Z. 1965. Zinacanteo "Souls." *Man* 65: 33–35.

Vogt, Evon Z. 1969. *Zinacantan: A Maya Community in the Highlands of Chiapas*. Cambridge, MA: Harvard University Press.

Vogt, Evon Z., and David Stuart. 2005. Some Notes on Ritual Caves among the Ancient and Modern Maya. In *In the Maw of the Earth Monster: Mesoamerican Ritual Cave Use*, edited by James E. Brady and Keith M. Prufer, pp. 155–185. Austin: University of Texas Press.

von Hagen, Andriana, and Craig Morris. 1998. *The Cities of the Ancient Andes*. London: Thames and Hudson.

Wasson, Gordon. 1980. *The Wondrous Mushroom*. New York: McGraw-Hill.

Whitten, Norman. 1976. *Sacha Runa: Ethnicity and Adaptation of the Ecuadorian Jungle Quichua*. Champaign-Urbana: University of Illinois Press.

Wichmann, Søren. 1995. The Relationship among the Mixe-Zoquean Languages of Mexico. Salt Lake City: University of Utah Press.

Wilbert, Johannes. 1987. *Tobacco Shamanism in South America*. New Haven, CT: Yale University Press.

Willey, Gordon R. 1955. The Prehistoric Civilizations of Nuclear America. *American Anthropologist* 57: 571–593.

Willey, Gordon R. 1962. The Great Art Styles and the Rise of Pre-Columbian Civilization. *American Anthropologist* 64 (1): 1–14.

Willey, Gordon R. 1999. Styles and State Formations. *Latin American Antiquity* 10 (1): 86–90.

Wilson, Richard. 1995. *Maya Resurgence in Guatemala*. Norman: University of Oklahoma Press.

Winter, Joseph C., ed. 2000. *Tobacco Use by Native North Americans: Sacred Smoke and Silent Killer*. Norman: University of Oklahoma Press.

Wisdom, Charles. 1940. *The Chorti Indians of Guatemala*. Chicago: University of Chicago Press.

Zuidema, R. Tom. 1964. *The Ceque System of Cuzco: The Social Organization of the Capital of the Inca*. International Archives of Ethnography, Supplement to Volume 50. Leiden: E. J. Brill.

Zuidema, R. Tom. 1973. Kinship and Ancestor Cult in Three Peruvian Communities: Hernández Príncipe's Account in 1622. *Bulletin, Institut français d'études andines* 2 (1): 16–23. Lima.

Zuidema, R. Tom. 1980. El *Ushnu*. Revista de la Universidad Complutense de Madrid. *Economía y Sociedad en los Andes y Mesoamérica* 28 (117): 317–362.

Zuidema, R. Tom. 1982. Myth and History in Ancient Peru. In *The Logic of Culture: Advances in Structural Theory and Methods*, edited by Ino Rossi, pp. 150–175. South Hadley, MA: J. F. Bergin.

Zuidema, R. Tom. 1983. Hierarchy and Space in Incaic Social Organization. *Ethnohistory* 30 (2): 49–75.

Zuidema, R. Tom. 1989. *Reyes y guerreros: Ensayos de cultura Andina*. Lima, Peru: Fomciencias.

Zuidema, R. Tom. 1992. Inca Cosmos in Andean Context: From the Perspective of Capac Raymi Carmay Quilla Feast Celebrating the December Solstice in Cuzco. In *Andean Cosmologies through Time: Persistence and Emergence*, edited by R. V. H. Dover, K. E. Seibold, and J. H. McDowell, pp. 17–45. Bloomington: University of Indiana Press.

Zuidema, R. Tom. 1995. *El sistema de ceques del Cuzco: La organización social de la capital del los Incas*. Con nuevo ensayo preliminar. Lima: Pontificia Universidad Católica del Peru. Fondo Editorial.

Zuidema, R. Tom. 1997. Cosmovisión Inca y astronomía en el Cuzco: Nuevo año agrícola y sucesión real. In *Pensar America Comovisión Mesoamerica y Andina*, edited by A. G. Aranda, pp. 249–270. Actas de las VI Jornadas del Inca Garcilaso celebrated in Montilla, September 11–13, 1996. Córdoba, Spain: Obra Social y Cutural Cajasur, Ayuntamiento de Montilla.

Zuidema, R. Tom. 2002. Inca Religion: Its Foundations in the Central Andean Context. In *Native Religions and Cultures of Central and South America*, edited by Lawrence E. Sullivan, pp. 236–253. London: Continuum.

Zuidema, R. Tom. 2008. The Astronomical Significance of Ritual Movements and in the Calendar of Cuzco. In *Pre-Columbian Landscapes of Creation and Origin*, edited by John E. Staller, pp. 249–268. New York: Springer.

Zuidema, R. Tom. 2011. *El calendario inca: Tiempo y espacio en la Organización Ritual del Cuzco; La idea del pasado*. Lima: Fondo Editorial del Congreso del Perú, Fondo Editorial Pontificia Universidad Católica del Perú.

Zumdahl, Steven S. 2009. *Chemical Principles*. 6th ed. New York: Houghton Mifflin.

INDEX